ALEKS Math
Study Guide
2020 - 2021

A Comprehensive Review and Step-By-Step Guide to Preparing for the ALEKS Math

By

Reza Nazari

All inquiries should be addressed to:

info@EffortlessMath.com

www.EffortlessMath.com

ISBN: 978-1-64612-869-3

Published by: **Effortless Math Education Inc.**

For Online Math Practice Visit www.EffortlessMath.com

Welcome to
ALEKS Math Prep
2021

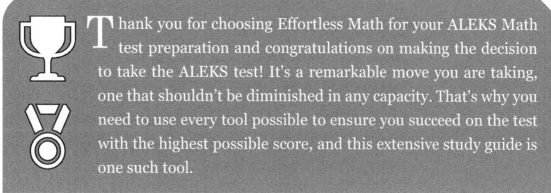

Thank you for choosing Effortless Math for your ALEKS Math test preparation and congratulations on making the decision to take the ALEKS test! It's a remarkable move you are taking, one that shouldn't be diminished in any capacity. That's why you need to use every tool possible to ensure you succeed on the test with the highest possible score, and this extensive study guide is one such tool.

If math has never been a strong subject for you, don't worry! This book will help you prepare for (and even ACE) the ALEKS Math assessment. As test day draws nearer, effective preparation becomes increasingly more important. Thankfully, you have this comprehensive study guide to help you get ready for the test. With this guide, you can feel confident that you will be more than ready for the ALEKS Math test when the time comes.

First and foremost, it is important to note that this book is a study guide and not a textbook. It is best read from cover to cover. Every lesson of this "self-guided math book" was carefully developed to ensure that you are making the most effective use of your time while preparing for the test. This up-to-date guide reflects the 2021 test guidelines and will put you on the right track to hone your math skills, overcome exam anxiety, and boost your confidence, so that you can have your best to succeed on the ALEKS Math test.

This study guide will:

☑ Explain the format of the ALEKS Math test.

☑ Describe specific test-taking strategies that you can use on the test.

☑ Provide ALEKS Math test-taking tips.

☑ Review all ALEKS Math concepts and topics you will be tested on.

☑ Help you identify the areas in which you need to concentrate your study time.

☑ Offer exercises that help you develop the basic math skills you will learn in each section.

☑ Give **2 realistic and full-length practice tests** (featuring new question types) with detailed answers to help you measure your exam readiness and build confidence.

This resource contains everything you will ever need to succeed on the ALEKS Math test. You'll get in-depth instructions on every math topic as well as tips and techniques on how to answer each question type. You'll also get plenty of practice questions to boost your test-taking confidence.

In addition, in the following pages you'll find:

➢ **How to Use This Book Effectively** – This section provides you with step-by-step instructions on how to get the most out of this comprehensive study guide.

➢ **How to study for the ALEKS Math Test** – A six-step study program has been developed to help you make the best use of this book and prepare for your ALEKS Math test. Here you'll find tips and strategies to guide your study program and help you understand ALEKS Math and how to ace the test.

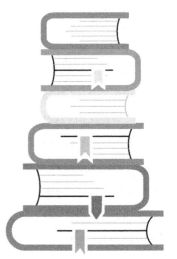

➢ **ALEKS Math Review** – Learn everything you need to know about the ALEKS Math test.

➢ **ALEKS Math Test-Taking Strategies** – Learn how to effectively put these recommended test-taking techniques into use for improving your ALEKS Math score.

➢ **Test Day Tips** – Review these tips to make sure you will do your best when the big day comes.

Effortless Math's ALEKS Online Center

Effortless Math Online ALEKS Center offers a complete study program, including the following:

✓ Step-by-step instructions on how to prepare for the ALEKS Math test

✓ Numerous ALEKS Math worksheets to help you measure your math skills

✓ Complete list of ALEKS Math formulas

✓ Video lessons for all ALEKS Math topics

✓ Full-length ALEKS Math practice tests

✓ And much more…

No Registration Required.

Visit **EffortlessMath.com/ALEKS** to find your online ALEKS Math resources.

How to Use This Book Effectively

L ook no further when you need a study guide to improve your math skills to succeed on the math portion of the ALEKS test. Each chapter of this comprehensive guide to the ALEKS Math will provide you with the knowledge, tools, and understanding needed for every topic covered on the test.

　　It's imperative that you understand each topic before moving onto another one, as that's the way to guarantee your success. Each chapter provides you with examples and a step-by-step guide of every concept to better understand the content that will be on the test. To get the best possible results from this book:

> **Begin studying long before your test date**. This provides you ample time to learn the different math concepts. The earlier you begin studying for the test, the sharper your skills will be. Do not procrastinate! Provide yourself with plenty of time to learn the concepts and feel comfortable that you understand them when your test date arrives.

> **Practice consistently**. Study ALEKS Math concepts at least 20 to 30 minutes a day. Remember, slow and steady wins the race, which can be applied to preparing for the ALEKS Math test. Instead of cramming to tackle everything at once, be patient and learn the math topics in short bursts.

> Whenever you get a math problem wrong, **mark it off, and review it later** to make sure you understand the concept.

> Start each session by looking over the previous related material.

> Once you've reviewed the book's lessons, **take a practice test** at the back of the book to gauge your level of readiness. Then, review your results. Read detailed answers and solutions for each question you missed.

> **Take another practice test** to get an idea of how ready you are to take the actual exam. Taking the practice tests will give you the confidence you need on test day. Simulate the ALEKS testing environment by sitting in a quiet room free from distraction. Make sure to clock yourself with a timer.

How to Study for the ALEKS Math Test

Studying for the ALEKS Math test can be a really daunting and boring task. What's the best way to go about it? Is there a certain study method that works better than others? Well, studying for the ALEKS Math can be done effectively. The following six-step program has been designed to make preparing for the ALEKS Math test more efficient and less overwhelming.

Step **1** - Create a study plan
Step **2** - Choose your study resources
Step **3** - Review, Learn, Practice
Step **4** - Learn and practice test-taking strategies
Step **5** - Learn the ALEKS Test format and take practice tests
Step **6** - Analyze your performance

STEP 1: Create a Study Plan

It's always easier to get things done when you have a plan. Creating a study plan for the ALEKS Math test can help you to stay on track with your studies. It's important to sit down and prepare a study plan with what works with your life, work, and any other obligations you may have. Devote enough time each day to studying. It's also a great idea to break down each section of the exam into blocks and study one concept at a time.

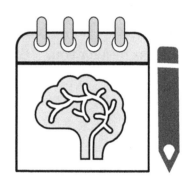

It's important to understand that there is no "right" way to create a study plan. Your study plan will be personalized based on your specific needs and learning style.

Follow these guidelines to create an effective study plan for your ALEKS Math test:

★ **Analyze your learning style and study habits** – Everyone has a different learning style. It is essential to embrace your individuality and the unique way you learn. Think about what works and what doesn't work for you. Do you prefer ALEKS Math prep books or a combination of textbooks and video lessons? Does it work better for you if you study every night for thirty minutes or is it more effective to study in the morning before going to work?

★ **Evaluate your schedule** – Review your current schedule and find out how much time you can consistently devote to ALEKS Math study.

★ **Develop a schedule** – Now it's time to add your study schedule to your calendar like any other obligation. Schedule time for study, practice, and review. Plan out which topic you will study on which day to ensure that you're devoting enough time to each concept. Develop a study plan that is mindful, realistic, and flexible.

★ **Stick to your schedule** – A study plan is only effective when it is followed consistently. You should try to develop a study plan that you can follow for the length of your study program.

★ **Evaluate your study plan and adjust as needed** – Sometimes you need to adjust your plan when you have new commitments. Check in with yourself regularly to make sure that you're not falling behind in your study plan. Remember, the most important thing is sticking to your plan. Your study plan is all about helping you be more productive. If you find that your study plan is not as effective as you want, don't get discouraged. It's okay to make changes as you figure out what works best for you.

STEP 2: Choose Your Study Resources

There are numerous textbooks and online resources available for the ALEKS Math test, and it may not be clear where to begin. Don't worry! This study guide provides everything you need to fully prepare for your ALEKS Math test. In addition to the book content, you can also use Effortless Math's online resources. (video lessons, worksheets, formulas, etc.)

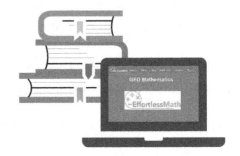

Simply visit EffortlessMath.com/ALEKS to find your online ALEKS Math resources.

STEP 3: Review, Learn, Practice

This ALEKS Math study guide breaks down each subject into specific skills or content areas. For instance, the percent concept is separated into different topics–percent calculation, percent increase and decrease, percent problems, etc. Use this book to help you go over all key math concepts and topics on the ALEKS Math test.

As you read each chapter, take notes or highlight the concepts you would like to go over again in the future. If you're unfamiliar with a topic or something is difficult for you, do additional research on it. For each math topic, plenty of instructions, step-by-step guides, and examples are provided to ensure you get a good grasp of the material. You can also find video lessons on the Effortless Math website for each ALEKS Math concept.

Quickly review the topics you do understand to get a brush-up of the material. Be sure to do the practice questions provided at the end of every chapter to measure your understanding of the concepts.

STEP 4: Learn and Practice Test-taking Strategies

In the following sections, you will find important test-taking strategies and tips that can help you earn extra points. You'll learn how to think strategically and when to guess if you don't know the answer to a question. Using ALEKS Math test-taking strategies and tips can help you raise your score and do well on the test. Apply test taking strategies on the practice tests to help you boost your confidence.

STEP 5: Learn the ALEKS Test Format and Take Practice Tests

The *ALEKS Test Review* section provides information about the structure of the ALEKS test. Read this section to learn more about the ALEKS test structure, different test sections, the number of questions in each section, and the section time limits. When you have a prior understanding of the test format and different types of ALEKS Math questions, you'll feel more confident when you take the actual exam.

Once you have read through the instructions and lessons and feel like you are ready to go – take advantage of both of the full-length ALEKS Math practice tests available in this study guide. Use the practice tests to sharpen your skills and build confidence.

The ALEKS Math practice tests offered at the end of the book are formatted similarly to the actual ALEKS Math test. When you take each practice test, try to simulate actual testing conditions. To take the practice tests, sit in a quiet space, time yourself, and work through as many of the questions as time allows. The practice tests are followed by detailed answer explanations to help you find your weak areas, learn from your mistakes, and raise your ALEKS Math score.

STEP 6: Analyze Your Performance

After taking the practice tests, look over the answer keys and explanations to learn which questions you answered correctly and which you did not. Never be discouraged if you make a few mistakes. See them as a learning opportunity. This will highlight your strengths and weaknesses.

You can use the results to determine if you need additional practice or if you are ready to take the actual ALEKS Math test.

Looking for more?

Visit EffortlessMath.com/ALEKS to find hundreds of ALEKS Math worksheets, video tutorials, practice tests, ALEKS Math formulas, and much more.

Or scan this QR code.

No Registration Required.

ALEKS Test Review

ALEKS (Assessment and Learning in Knowledge Spaces) is an artificial intelligence-based assessment tool to measure the strengths and weaknesses of a student's mathematical knowledge. ALEKS is available for a variety of subjects and courses in K-12, Higher Education, and Continuing Education. The findings of ALEKS's assessment test help to find an appropriate level for course placement. The ALEKS math placement assessment ensures students' readiness for particular math courses at colleges.

ALEKS does not use multiple-choice questions like most other standardized tests. Instead, it utilizes adaptable and easy-to-use method that mimic paper and pencil techniques. When taking the ALEKS test, a brief tutorial helps you learn how to use ALEKS answer input tools. You then begin the ALEKS Assessment. In about 30 to 45 minutes, the test measures your current content knowledge by asking 20 to 30 questions. ALEKS is a Computer Adaptive (CA) assessment. It means that each question will be chosen on the basis of answers to all the previous questions. Therefore, each set of assessment questions is unique. The ALEKS Math assessment does not allow you to use a personal calculator. But for some questions ALEKS onscreen calculator button is active and the test taker can use it.

Key Features of the ALEKS Mathematics Assessment

Some key features of the ALEKS Math assessment are:

- ❖ Mathematics questions on ALEKS are adaptive to identify the student's knowledge from a comprehensive standard curriculum, ranging from basic arithmetic up to precalculus, including trigonometry but not calculus.

- ❖ Unlike other standardized tests, the ALEKS assessment does not provide a "grade" or "raw score." Instead, ALEKS identifies which concepts the student has mastered and what topics the student needs to learn.

- ❖ ALEKS does not use multiple-choice questions. Instead, students need to produce authentic mathematical input.

- ❖ There is no time limit for taking the ALEKS Math assessment. But it usually takes 30 to 45 minutes to complete the assessment.

The ALEKS Math score is between 1 and 100 and is interpreted as a percentage correct. A higher ALEKS score indicates that the test-taker has mastered more math concepts. ALEKS Math assessment tool evaluates mastery of a comprehensive set of mathematics skills ranging from basic arithmetic up to precalculus, including trigonometry but not calculus. It will place students in classes up to Calculus.

ALEKS Math Test-Taking Strategies

Here are some test-taking strategies that you can use to maximize your performance and results on the ALEKS Math test.

#1: USE THIS APPROACH TO ANSWER EVERY ALEKS MATH QUESTION

- Review the question to identify keywords and important information.

- Translate the keywords into math operations so you can solve the problem.

- Review the answer choices. What are the differences between answer choices?

- Draw or label a diagram if needed.

- Try to find patterns.

- Find the right method to answer the question. Use straightforward math, plug in numbers, or test the answer choices (backsolving).

- Double-check your work.

#2: ANSWER EVERY ALEKS MATH QUESTION

Don't leave any fields empty! ALEKS is a Computer Adaptive (CA) assessment. Therefore, you cannot leave a question unanswered and you cannot go back to previous questions.

Even if you're unable to work out a problem, strive to answer it. Take a guess if you have to. You will not lose points by getting an answer wrong, though you may gain a point by getting it correct!

#3: BALLPARK

A ballpark answer is a rough approximation. When we become overwhelmed by calculations and figures, we end up making silly mistakes. A decimal that is moved by one unit can change an answer from right to wrong, regardless of the number of steps that you went through to get it. That's where ballparking can play a big part.

If you think you know what the correct answer may be (even if it's just a ballpark answer), you'll usually have the ability to estimate the range of possible answers and avoid simple mistakes.

#4: PLUGGING IN NUMBERS

"Plugging in numbers" is a strategy that can be applied to a wide range of different math problems on the ALEKS Math test. This approach is typically used to simplify a challenging question so that it is more understandable. By using the strategy carefully, you can find the answer without too much trouble.

The concept is fairly straightforward–replace unknown variables in a problem with certain values. When selecting a number, consider the following:

- Choose a number that's basic (just not too basic). Generally, you should avoid choosing 1 (or even 0). A decent choice is 2.

- Try not to choose a number that is displayed in the problem.

- Make sure you keep your numbers different if you need to choose at least two of them.

- If your question contains fractions, then a potential right answer may involve either an LCD (least common denominator) or an LCD multiple.

- 100 is the number you should choose when you are dealing with problems involving percentages.

ALEKS Mathematics – Test Day Tips

After practicing and reviewing all the math concepts you've been taught, and taking some ALEKS mathematics practice tests, you'll be prepared for test day. Consider the following tips to be extra-ready come test time.

Before Your Test

What to do the night before:

- **Relax!** One day before your test, study lightly or skip studying altogether. You shouldn't attempt to learn something new, either. There are plenty of reasons why studying the evening before a big test can work against you. Put it this way–a marathoner wouldn't go out for a sprint before the day of a big race. Mental marathoners–such as yourself–should not study for any more than one hour 24 hours before a ALEKS test. That's because your brain requires some rest to be at its best. The night before your exam, spend some time with family or friends, or read a book.

- **Avoid bright screens** - You'll have to get some good shuteye the night before your test. Bright screens (such as the ones coming from your laptop, TV, or mobile device) should be avoided altogether. Staring at such a screen will keep your brain up, making it hard to drift asleep at a reasonable hour.

- **Make sure your dinner is healthy** - The meal that you have for dinner should be nutritious. Be sure to drink plenty of water as well. Load up on your complex carbohydrates, much like a marathon runner would do. Pasta, rice, and potatoes are ideal options here, as are vegetables and protein sources.

- **Get your bag ready for test day** – Prefer to take ALEKS in the Testing Office? The night prior to your test, pack your bag with your stationery, admissions pass, ID, and any other gear that you need. Keep the bag right by your front door. If you prefer to take the test at home, find a quite place without any distractions.

- **Make plans to reach the testing site** – If you are taking the test at the testing office, ensure that you understand precisely how you will arrive at the site of the test. If parking is something you'll have to find first, plan for it. If you're dependent on public transit, then review the schedule. You should also make sure that the train/bus/subway/streetcar you use will be running. Find out about road closures as well. If a parent or friend is accompanying you, ensure that they understand what steps they have to take as well.

The Day of the Test

- Get up reasonably early, but not too early.

- Have breakfast - Breakfast improves your concentration, memory, and mood. As such, make sure the breakfast that you eat in the morning is healthy. The last thing you want to be is distracted by a grumbling tummy. If it's not your own stomach making those noises, another test taker close to you might be instead. Prevent discomfort or embarrassment by consuming a healthy breakfast. Bring a snack with you if you think you'll need it.

- Follow your daily routine - Do you watch TV in the morning while getting ready for the day? Don't break your usual habits on the day of the test. Likewise, if coffee isn't something you drink in the morning, then don't take up the habit hours before your test. Routine consistency lets you concentrate on the main objective–doing the best you can on your test.

- Wear layers - Dress yourself up in comfortable layers if you are taking the test at the testing site. You should be ready for any kind of internal temperature. If it gets too warm during the test, take a layer off.

- Make your voice heard - If something is off, speak to a proctor. If medical attention is needed or if you'll require anything, consult the proctor prior to the start of the test. Any doubts you have should be clarified. You should be entering the test site with a state of mind that is completely clear.

- Have faith in yourself - When you feel confident, you will be able to perform at your best. When you are waiting for the test to begin, envision yourself receiving an outstanding result. Try to see yourself as someone who knows all the answers, no matter what the questions are. A lot of athletes tend to use this technique–particularly before a big competition. Your expectations will be reflected by your performance.

During your test

- Be calm and breathe deeply - You need to relax before the test, and some deep breathing will go a long way to help you do that. Be confident and calm. You got this. Everybody feels a little stressed out just before an evaluation of any kind is set to begin. Learn some effective breathing exercises. Spend a minute meditating before the test starts. Filter out any negative thoughts you have. Exhibit confidence when having such thoughts.

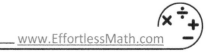

- Concentrate on the test - Refrain from comparing yourself to anyone else. You shouldn't be distracted by the people near you or random noise. Concentrate exclusively on the test. If you find yourself irritated by surrounding noises, earplugs can be used to block sounds off close to you. Don't forget–the test is going to last an hour or more. Some of that time will be dedicated to brief sections. Concentrate on the specific section you are working on during a particular moment. Do not let your mind wander off to upcoming or previous questions.

- Try to answer each question individually - Focus only on the question you are working on. Use one of the test-taking strategies to solve the problem. If you aren't able to come up with an answer, don't get frustrated. Simply guess, then move onto the next question.

- Don't forget to breathe! Whenever you notice your mind wandering, your stress levels boosting, or frustration brewing, take a thirty-second break. Shut your eyes, drop your pencil, breathe deeply, and let your shoulders relax. You will end up being more productive when you allow yourself to relax for a moment.

After your test

- Take it easy - You will need to set some time aside to relax and decompress once the test has concluded. There is no need to stress yourself out about what you could've said, or what you may have done wrong. At this point, there's nothing you can do about it. Your energy and time would be better spent on something that will bring you happiness for the remainder of your day.

- Redoing the test - Did you succeed on the test? Congratulations! Your hard work paid off! Succeeding on this test means that you are now ready to take college level courses.

  If you didn't receive the result you expected, though, don't worry! The test can be retaken. In such cases, you will need to follow the retake policy. You also need to re-register to take the exam again.

Contents

 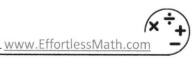

Name: ....

Date: ....

Topic	Simplifying Fractions
Notes	✓ Evenly divide both the top and bottom of the fraction by 2, 3, 5, 7, ... etc. ✓ Continue until you can't go any further.
Example	**Simplify** $\frac{36}{48}$ To simplify $\frac{36}{48}$, find a number that both 36 and 48 are divisible by. Both are divisible by 12. Then: $\frac{36}{48} = \frac{36 \div 12}{48 \div 12} = \frac{3}{4}$

Your Turn!		
	1) $\frac{2}{18} =$	2) $\frac{22}{66} =$
	3) $\frac{12}{48} =$	4) $\frac{11}{99} =$
	5) $\frac{15}{75} =$	6) $\frac{25}{100} =$
	7) $\frac{16}{72} =$	8) $\frac{32}{96} =$
	9) $\frac{14}{77} =$	10) $\frac{60}{84} =$

Name:	Date: ....

Topic	**Simplifying Fractions - Answers**
Notes	✓ Evenly divide both the top and bottom of the fraction by $2, 3, 5, 7, \dots$ etc. ✓ Continue until you can't go any further.
Example	**Simplify** $\frac{36}{48}$ To simplify $\frac{36}{48}$, find a number that both 36 and 48 are divisible by. Both are divisible by 12. Then: $\frac{36}{48} = \frac{36 \div 12}{48 \div 12} = \frac{3}{4}$

Your Turn!	1) $\frac{2}{18} = \frac{1}{9}$	2) $\frac{22}{66} = \frac{1}{3}$
	3) $\frac{12}{48} = \frac{1}{4}$	4) $\frac{11}{99} = \frac{1}{9}$
	5) $\frac{15}{75} = \frac{1}{5}$	6) $\frac{25}{100} = \frac{1}{4}$
	7) $\frac{16}{72} = \frac{2}{9}$	8) $\frac{32}{96} = \frac{1}{3}$
	9) $\frac{14}{77} = \frac{2}{11}$	10) $\frac{60}{84} = \frac{5}{7}$

Name:	Date: ....

Topic	**Adding and Subtracting Fractions**
Notes	✓ For "like" fractions (fractions with the same denominator), add or subtract the numerators and write the answer over the common denominator. ✓ Find equivalent fractions with the same denominator before you can add or subtract fractions with different denominators. ✓ Adding and Subtracting with the same denominator: $\frac{a}{b}+\frac{c}{b}=\frac{a+c}{b}, \frac{a}{b}-\frac{c}{b}=\frac{a-c}{b}$ ✓ Adding and Subtracting fractions with different denominators: $\frac{a}{b}+\frac{c}{d}=\frac{ad+bc}{bd}, \frac{a}{b}-\frac{c}{d}=\frac{ad-bc}{bd}$
Example	**Find the sum.** $\frac{3}{5}+\frac{2}{3}=\frac{(3)3+(5)(2)}{5\times 3}=\frac{19}{15}$ **Subtract.** $\frac{4}{7}-\frac{3}{7}=\frac{1}{7}$

Your Turn!	1) $\frac{3}{5}+\frac{2}{7}=$	2) $\frac{7}{9}-\frac{4}{7}=$
	3) $\frac{4}{9}+\frac{5}{8}=$	4) $\frac{5}{8}-\frac{2}{5}=$
	5) $\frac{2}{5}+\frac{1}{6}=$	6) $\frac{2}{3}-\frac{1}{4}=$
	7) $\frac{8}{9}+\frac{5}{7}=$	8) $\frac{6}{7}-\frac{5}{9}=$

Name:	Date: ....

Topic	**Adding and Subtracting Fractions - Answers**
Notes	✓ For "like" fractions (fractions with the same denominator), add or subtract the numerators and write the answer over the common denominator. ✓ Find equivalent fractions with the same denominator before you can add or subtract fractions with different denominators. ✓ Adding and Subtracting with the same denominator: $$\frac{a}{b}+\frac{c}{b}=\frac{a+c}{b}, \frac{a}{b}-\frac{c}{b}=\frac{a-c}{b}$$ ✓ Adding and Subtracting fractions with different denominators: $$\frac{a}{b}+\frac{c}{d}=\frac{ad+bc}{bd}, \frac{a}{b}-\frac{c}{d}=\frac{ad-bc}{bd}$$
Example	**Find the sum.** $\dfrac{3}{5}+\dfrac{2}{3}=\dfrac{(3)3+(5)(2)}{5\times3}=\dfrac{19}{15}$ **Subtract.** $\dfrac{4}{7}-\dfrac{3}{7}=\dfrac{1}{7}$
Your Turn!	1) $\dfrac{3}{5}+\dfrac{2}{7}=\dfrac{31}{35}$ 2) $\dfrac{7}{9}-\dfrac{4}{7}=\dfrac{13}{63}$ 3) $\dfrac{4}{9}+\dfrac{5}{8}=\dfrac{77}{72}$ 4) $\dfrac{5}{8}-\dfrac{2}{5}=\dfrac{9}{40}$ 5) $\dfrac{2}{5}+\dfrac{1}{6}=\dfrac{17}{30}$ 6) $\dfrac{2}{3}-\dfrac{1}{4}=\dfrac{5}{12}$ 7) $\dfrac{8}{9}+\dfrac{5}{7}=\dfrac{101}{63}$ 8) $\dfrac{6}{7}-\dfrac{5}{9}=\dfrac{19}{63}$

Name:	Date: ....

Topic	**Multiplying and Dividing Fractions**
Notes	✓ Multiplying fractions: multiply the top numbers and multiply the bottom numbers. ✓ Dividing fractions: Keep, Change, Flip Keep first fraction, change division sign to multiplication, and flip the numerator and denominator of the second fraction. Then, solve!
Examples	**Multiply.** $\frac{2}{5} \times \frac{3}{4} =$ Multiply the top numbers and multiply the bottom numbers. $\frac{2}{5} \times \frac{3}{4} = \frac{2 \times 3}{5 \times 4} = \frac{6}{20}$, simplify: $\frac{6}{20} = \frac{6 \div 2}{20 \div 2} = \frac{3}{10}$ **Divide.** $\frac{2}{5} \div \frac{3}{4} =$ Keep first fraction, change division sign to multiplication, and flip the numerator and denominator of the second fraction. Then: $\frac{2}{5} \div \frac{3}{4} = \frac{2}{5} \times \frac{4}{3} = \frac{2 \times 4}{5 \times 3} = \frac{8}{15}$
Your Turn!	1) $\frac{5}{9} \times \frac{4}{7} =$ 2) $\frac{3}{5} \div \frac{2}{3} =$ 3) $\frac{2}{7} \times \frac{3}{5} =$ 4) $\frac{2}{5} \div \frac{7}{12} =$ 5) $\frac{1}{7} \times \frac{4}{9} =$ 6) $\frac{2}{9} \div \frac{3}{7} =$ 7) $\frac{2}{5} \times \frac{6}{7} =$ 8) $\frac{1}{4} \div \frac{2}{5} =$

Name:	Date: ....

Topic	**Multiplying and Dividing Fractions - Answers**
Notes	✓ Multiplying fractions: multiply the top numbers and multiply the bottom numbers. ✓ Dividing fractions: Keep, Change, Flip Keep first fraction, change division sign to multiplication, and flip the numerator and denominator of the second fraction. Then, solve!
Examples	**Multiply.** $\frac{2}{5} \times \frac{3}{4} =$ Multiply the top numbers and multiply the bottom numbers. $\frac{2}{5} \times \frac{3}{4} = \frac{2 \times 3}{5 \times 4} = \frac{6}{20}$, simplify: $\frac{6}{20} = \frac{6 \div 2}{20 \div 2} = \frac{3}{10}$ **Divide.** $\frac{2}{5} \div \frac{3}{4} =$ Keep first fraction, change division sign to multiplication, and flip the numerator and denominator of the second fraction. Then: $\frac{2}{5} \div \frac{3}{4} = \frac{2}{5} \times \frac{4}{3} = \frac{2 \times 4}{5 \times 3} = \frac{8}{15}$

Your Turn!		
	1) $\frac{5}{9} \times \frac{4}{7} = \frac{20}{63}$	2) $\frac{3}{5} \div \frac{2}{3} = \frac{9}{10}$
	3) $\frac{2}{7} \times \frac{3}{5} = \frac{6}{35}$	4) $\frac{2}{5} \div \frac{7}{12} = \frac{24}{35}$
	5) $\frac{1}{7} \times \frac{4}{9} = \frac{4}{63}$	6) $\frac{2}{9} \div \frac{3}{7} = \frac{14}{27}$
	7) $\frac{2}{5} \times \frac{6}{7} = \frac{12}{35}$	8) $\frac{1}{4} \div \frac{2}{5} = \frac{5}{8}$

Name:	Date: ....

Topic	**Adding Mixed Numbers**
Notes	Use the following steps for adding mixed numbers. ✓ Add whole numbers of the mixed numbers. ✓ Add the fractions of each mixed number. ✓ Find the Least Common Denominator (LCD) if necessary. ✓ Add whole numbers and fractions. ✓ Write your answer in lowest terms.
Example	**Add mixed numbers.** $1\frac{1}{2} + 2\frac{2}{3} =$ Rewriting our equation with parts separated, $1 + \frac{1}{2} + 2 + \frac{2}{3}$ Add whole numbers: $1 + 2 = 3$ Add fractions: $\frac{1}{2} + \frac{2}{3} = \frac{3}{6} + \frac{4}{6} = \frac{7}{6} = 1\frac{1}{6}$ Now, combine the whole and fraction parts: $3 + 1 + \frac{1}{6} = 4\frac{1}{6}$
Your Turn!	1) $1\frac{1}{12} + 2\frac{3}{4} =$ 2) $3\frac{5}{8} + 1\frac{1}{4} =$ 3) $1\frac{1}{10} + 2\frac{2}{5} =$ 4) $2\frac{5}{6} + 2\frac{2}{9} =$ 5) $2\frac{2}{7} + 1\frac{2}{21} =$ 6) $1\frac{3}{8} + 3\frac{2}{3} =$ 7) $3\frac{1}{5} + 1\frac{2}{8} =$ 8) $3\frac{1}{2} + 2\frac{3}{7} =$

Name:	Date: ....

Topic	**Adding Mixed Numbers - Answers**
Notes	Use the following steps for adding mixed numbers. ✓ Add whole numbers of the mixed numbers. ✓ Add the fractions of each mixed number. ✓ Find the Least Common Denominator (LCD) if necessary. ✓ Add whole numbers and fractions. ✓ Write your answer in lowest terms.
Example	**Add mixed numbers.** $1\frac{1}{2} + 2\frac{2}{3} =$ Rewriting our equation with parts separated, $1 + \frac{1}{2} + 2 + \frac{2}{3}$ Add whole numbers: $1 + 2 = 3$ Add fractions: $\frac{1}{2} + \frac{2}{3} = \frac{3}{6} + \frac{4}{6} = \frac{7}{6} = 1\frac{1}{6}$ Now, combine the whole and fraction parts: $3 + 1 + \frac{1}{6} = 4\frac{1}{6}$

Your Turn!	1) $1\frac{1}{12} + 2\frac{3}{4} = 3\frac{5}{6}$	2) $3\frac{5}{8} + 1\frac{1}{4} = 4\frac{7}{8}$
	3) $1\frac{1}{10} + 2\frac{2}{5} = 3\frac{1}{2}$	4) $2\frac{5}{6} + 2\frac{2}{9} = 5\frac{1}{18}$
	5) $2\frac{2}{7} + 1\frac{2}{21} = 3\frac{8}{21}$	6) $1\frac{3}{8} + 3\frac{2}{3} = 5\frac{1}{24}$
	7) $3\frac{1}{5} + 1\frac{2}{8} = 4\frac{9}{20}$	8) $3\frac{1}{2} + 2\frac{3}{7} = 5\frac{13}{14}$

Name:	Date: ....

Topic	**Subtracting Mixed Numbers**
Notes	Use the following steps for subtracting mixed numbers. ✓ Convert mixed numbers into improper fractions. $a\frac{c}{b} = \frac{ab+c}{b}$ ✓ Find equivalent fractions with the same denominator for unlike fractions (fractions with different denominators) ✓ Subtract the second fraction from the first one. ✓ Write your answer in lowest terms and convert it into a mixed number if the answer is an improper fraction.
Example	**Subtract.** $5\frac{1}{2} - 2\frac{2}{3} =$ Convert mixed numbers into fractions: $5\frac{1}{2} = \frac{5 \times 2 + 1}{5} = \frac{11}{2}$ and $2\frac{2}{3} = \frac{2 \times 3 + 2}{4} = \frac{8}{3}$, these two fractions are "unlike" fractions. (they have different denominators). Find equivalent fractions with the same denominator. Use this formula: $\frac{a}{b} - \frac{c}{d} = \frac{ad - bc}{bd}$ $\frac{11}{2} - \frac{8}{3} = \frac{(11)(3) - (2)(8)}{2 \times 3} = \frac{33 - 16}{6} = \frac{17}{6}$, the answer is an improper fraction, convert it into a mixed number. $\frac{17}{6} = 2\frac{5}{6}$
Your Turn!	1) $2\frac{2}{5} - 1\frac{1}{3} =$ 2) $3\frac{5}{8} - 2\frac{1}{3} =$ 3) $6\frac{1}{4} - 1\frac{2}{7} =$ 4) $8\frac{2}{3} - 1\frac{1}{4} =$ 5) $8\frac{3}{4} - 1\frac{3}{8} =$ 6) $2\frac{3}{8} - 1\frac{2}{3} =$ 7) $13\frac{2}{7} - 1\frac{2}{21} =$ 8) $5\frac{1}{2} - 2\frac{3}{7} =$

| Name: | .... | Date: | .... |

Topic	**Subtracting Mixed Numbers - Answers**
Notes	Use the following steps for subtracting mixed numbers. ✓ Convert mixed numbers into improper fractions. $a\frac{c}{b} = \frac{ab+c}{b}$ ✓ Find equivalent fractions with the same denominator for unlike fractions (fractions with different denominators) ✓ Subtract the second fraction from the first one. ✓ Write your answer in lowest terms and convert it into a mixed number if the answer is an improper fraction.
Example	**Subtract.** $5\frac{1}{2} - 2\frac{2}{3} =$ Convert mixed numbers into fractions: $5\frac{1}{2} = \frac{5 \times 2 + 1}{5} = \frac{11}{2}$ and $2\frac{2}{3} = \frac{2 \times 3 + 2}{4} = \frac{8}{3}$, these two fractions are "unlike" fractions. (they have different denominators). Find equivalent fractions with the same denominator. Use this formula: $\frac{a}{b} - \frac{c}{d} = \frac{ad - bc}{bd}$ $\frac{11}{2} - \frac{8}{3} = \frac{(11)(3) - (2)(8)}{2 \times 3} = \frac{33 - 16}{6} = \frac{17}{6}$, the answer is an improper fraction, convert it into a mixed number. $\frac{17}{6} = 2\frac{5}{6}$
Your Turn!	1) $2\frac{2}{5} - 1\frac{1}{3} = 1\frac{1}{15}$ 2) $3\frac{5}{8} - 2\frac{1}{3} = 1\frac{7}{24}$ 3) $6\frac{1}{4} - 1\frac{2}{7} = 4\frac{27}{28}$ 4) $8\frac{2}{3} - 1\frac{1}{4} = 7\frac{5}{12}$ 5) $8\frac{3}{4} - 1\frac{3}{8} = 7\frac{3}{8}$ 6) $2\frac{3}{8} - 1\frac{2}{3} = \frac{17}{24}$ 7) $13\frac{2}{7} - 1\frac{2}{21} = 12\frac{4}{21}$ 8) $5\frac{1}{2} - 2\frac{3}{7} = 3\frac{1}{14}$

Name:	Date: ....

Topic	**Multiplying Mixed Numbers**	
Notes	✓ Convert the mixed numbers into fractions. $a\frac{c}{b} = a + \frac{c}{b} = \frac{ab+c}{b}$ ✓ Multiply fractions and simplify if necessary. $\frac{a}{b} \times \frac{c}{d} = \frac{a \times c}{b \times d}$ ✓ If the answer is an improper fraction (numerator is bigger than denominator), convert it into a mixed number.	
Example	**Multiply** $2\frac{1}{4} \times 3\frac{1}{2}$ Convert mixed numbers into fractions: $2\frac{1}{4} = \frac{2\times4+1}{4} = \frac{9}{4}$ and $3\frac{1}{2} = \frac{3\times2+1}{2} = \frac{7}{2}$ Multiply two fractions: $\frac{9}{4} \times \frac{7}{2} = \frac{9\times7}{4\times2} = \frac{63}{8}$ The answer is an improper fraction. Convert it into a mixed number: $$\frac{63}{8} = 7\frac{7}{8}$$	
Your Turn!	1) $5\frac{2}{3} \times 2\frac{2}{9} =$	2) $4\frac{1}{6} \times 5\frac{3}{7} =$
	3) $3\frac{1}{3} \times 3\frac{3}{4} =$	4) $2\frac{2}{9} \times 6\frac{1}{3} =$
	5) $2\frac{2}{7} \times 4\frac{3}{5} =$	6) $1\frac{4}{7} \times 9\frac{1}{2} =$
	7) $4\frac{1}{8} \times 3\frac{2}{3} =$	8) $6\frac{2}{3} \times 1\frac{1}{4} =$

Name:	Date: ....

Topic	**Multiplying Mixed Numbers - Answers**
Notes	✓ Convert the mixed numbers into fractions. $a\frac{c}{b} = a + \frac{c}{b} = \frac{ab+c}{b}$ ✓ Multiply fractions and simplify if necessary. $\frac{a}{b} \times \frac{c}{d} = \frac{a \times c}{b \times d}$ ✓ If the answer is an improper fraction (numerator is bigger than denominator), convert it into a mixed number.
Example	**Multiply** $2\frac{1}{4} \times 3\frac{1}{2}$ Convert mixed numbers into fractions: $2\frac{1}{4} = \frac{2\times4+1}{4} = \frac{9}{4}$ and $3\frac{1}{2} = \frac{3\times2+1}{2} = \frac{7}{2}$ Multiply two fractions: $\frac{9}{4} \times \frac{7}{2} = \frac{9\times7}{4\times2} = \frac{63}{8}$ The answer is an improper fraction. Convert it into a mixed number: $$\frac{63}{8} = 7\frac{7}{8}$$
Your Turn!	1) $5\frac{2}{3} \times 2\frac{2}{9} = 12\frac{16}{27}$ 2) $4\frac{1}{6} \times 5\frac{3}{7} = 22\frac{13}{21}$ 3) $3\frac{1}{3} \times 3\frac{3}{4} = 12\frac{1}{2}$ 4) $2\frac{2}{9} \times 6\frac{1}{3} = 14\frac{2}{27}$ 5) $2\frac{2}{7} \times 4\frac{3}{5} = 10\frac{18}{35}$ 6) $1\frac{4}{7} \times 9\frac{1}{2} = 14\frac{13}{14}$ 7) $4\frac{1}{8} \times 3\frac{2}{3} = 15\frac{1}{8}$ 8) $6\frac{2}{3} \times 1\frac{1}{4} = 8\frac{1}{3}$

Name:	Date: ....

Topic	Dividing Mixed Numbers
Notes	✓ Convert the mixed numbers into improper fractions. $$a\frac{c}{b} = a + \frac{c}{b} = \frac{ab+c}{b}$$ ✓ Divide fractions and simplify if necessary.
Example	**Solve.** $2\frac{1}{3} \div 1\frac{1}{4} =$ Converting mixed numbers to fractions: $2\frac{1}{3} \div 1\frac{1}{4} = \frac{7}{3} \div \frac{5}{4}$ Keep, Change, Flip: $\frac{7}{3} \div \frac{5}{4} = \frac{7}{3} \times \frac{4}{5} = \frac{7\times4}{3\times5} = \frac{28}{15} = 1\frac{13}{15}$
Your Turn!	1) $3\frac{2}{7} \div 2\frac{1}{4} =$ 2) $4\frac{2}{9} \div 1\frac{5}{6} =$ 3) $4\frac{2}{3} \div 3\frac{2}{5} =$ 4) $5\frac{4}{5} \div 4\frac{3}{4} =$ 5) $1\frac{8}{9} \div 2\frac{3}{7} =$ 6) $3\frac{3}{8} \div 2\frac{2}{5} =$ 7) $4\frac{1}{5} \div 3\frac{1}{9} =$ 8) $4\frac{2}{3} \div 1\frac{8}{9} =$ 9) $5\frac{2}{3} \div 3\frac{3}{7} =$ 10) $7\frac{1}{2} \div 5\frac{1}{3} =$

Name:	Date: ....

Topic	**Dividing Mixed Numbers- Answers**
Notes	✓ Convert the mixed numbers into improper fractions. $$a\frac{c}{b} = a + \frac{c}{b} = \frac{ab+c}{b}$$ ✓ Divide fractions and simplify if necessary.
Example	**Solve.** $2\frac{1}{3} \div 1\frac{1}{4} =$ Converting mixed numbers to fractions: $2\frac{1}{3} \div 1\frac{1}{4} = \frac{7}{3} \div \frac{5}{4}$ Keep, Change, Flip: $\frac{7}{3} \div \frac{5}{4} = \frac{7}{3} \times \frac{4}{5} = \frac{7\times4}{3\times5} = \frac{28}{15} = 1\frac{13}{15}$
Your Turn!	1) $3\frac{2}{7} \div 2\frac{1}{4} = 1\frac{29}{63}$ 2) $4\frac{2}{9} \div 1\frac{5}{6} = 2\frac{10}{33}$ 3) $4\frac{2}{3} \div 3\frac{2}{5} = 1\frac{19}{51}$ 4) $5\frac{4}{5} \div 4\frac{3}{4} = 1\frac{21}{95}$ 5) $1\frac{8}{9} \div 2\frac{3}{7} = \frac{7}{9}$ 6) $3\frac{3}{8} \div 2\frac{2}{5} = 1\frac{13}{32}$ 7) $4\frac{1}{5} \div 3\frac{1}{9} = 1\frac{7}{20}$ 8) $4\frac{2}{3} \div 1\frac{8}{9} = 2\frac{8}{17}$ 9) $5\frac{2}{3} \div 3\frac{3}{7} = 1\frac{47}{72}$ 10) $7\frac{1}{2} \div 5\frac{1}{3} = 1\frac{13}{32}$

Name:	Date: ....

Topic	Comparing Decimals
Notes	Decimals: is a fraction written in a special form. For example, instead of writing $\frac{1}{2}$ you can write 0.5. For comparing decimals: ✓ Compare each digit of two decimals in the same place value. ✓ Start from left. Compare hundreds, tens, ones, tenth, hundredth, etc. ✓ To compare numbers, use these symbols: - Equal to =, Less than <, Greater than > Greater than or equal ≥, Less than or equal ≤

Examples	**Compare 0.40 and 0.04.**
	
	  0.40 is greater than 0.04, because the tenth place of 0.40 is 4, but the tenth place of 0.04 is zero. Then: 0.40 > 0.04
	
	 Compare 0.0912 and 0.912.
	
	  0.912 is greater than 0.0912, because the tenth place of 0.912 is 9, but the tenth place of 0.0912 is zero. Then: 0.0912 < 0.912

Your Turn!	1) 0.91 ☐ 0.95	2) 1.79 ☐ 1.80
	3) 19.1 ☐ 19.09	4) 2.45 ☐ 2.089
	5) 1.258 ☐ 12.58	6) 0.89 ☐ 0.890
	7) 3.871 ☐ 2.998	8) 0.567 ☐ 0.756

Name:	Date: ....

Topic	**Comparing Decimals - Answers**
Notes	Decimals: is a fraction written in a special form. For example, instead of writing $\frac{1}{2}$ you can write 0.5. For comparing decimals: ✓ Compare each digit of two decimals in the same place value. ✓ Start from left. Compare hundreds, tens, ones, tenth, hundredth, etc. ✓ To compare numbers, use these symbols: - Equal to =, Less than <, Greater than > Greater than or equal ≥, Less than or equal ≤

Examples	**Compare 0.40 and 0.04.**
	
	    0.40 is greater than 0.04, because the tenth place of 0.40 is 4, but the tenth place of 0.04 is zero. Then: $0.40 > 0.04$
	
	 Compare 0.0912 and 0.912.
	
	    0.912 is greater than 0.0912, because the tenth place of 0.912 is 9, but the tenth place of 0.0912 is zero. Then: $0.0912 < 0.912$

Your Turn!	1) $0.91 < 0.95$	2) $1.79 < 1.80$
	3) $19.1 > 19.09$	4) $2.45 > 2.089$
	5) $1.258 < 12.58$	6) $0.89 = 0.890$
	7) $3.387 > 2.998$	8) $0.567 < 0.756$

Name:	Date: ....

Topic	**Rounding Decimals**
Notes	✓ We can round decimals to a certain accuracy or number of decimal places. ✓ Let's review place values: For example: 35.4817 3: tens 5: ones 4: tenths 8: hundredths 1: thousandths 7: tens thousandths ✓ To round a decimal, find the place value you'll round to. ✓ Find the digit to the right of the place value you're rounding to. If it is 5 or bigger, add 1 to the place value you're rounding to and remove all digits on its right side. If the digit to the right of the place value is less than 5, keep the place value and remove all digits on the right.
Example	**Round 12.8365 to the hundredth-place value.** First look at the next place value to the right, (thousandths). It's 6 and it is greater than 5. Thus add 1 to the digit in the hundredth place. It is 3. → $3 + 1 = 4$, then, the answer is 12.84
Your Turn!	**Round each number to the underlined place value.** 1) $32.5\underline{4}8 =$ | 2) $2.3\underline{2}6 =$ 3) $55.\underline{4}23 =$ | 4) $25.\underline{6}2 =$ 5) $11.\underline{2}65 =$ | 6) $33.5\underline{0}5 =$ 7) $3.5\underline{8}9 =$ | 8) $8.0\underline{1}9 =$

| Name: .... | Date: .... |

Topic	**Rounding Decimals - Answers**
Notes	✓ We can round decimals to a certain accuracy or number of decimal places. ✓ Let's review place values: For example: 35.4817 3: tens 5: ones 4: tenths 8: hundredths 1: thousandths 7: tens thousandths ✓ To round a decimal, find the place value you'll round to. ✓ Find the digit to the right of the place value you're rounding to. If it is 5 or bigger, add 1 to the place value you're rounding to and remove all digits on its right side. If the digit to the right of the place value is less than 5, keep the place value and remove all digits on the right.
Example	**Round 12.8365 to the hundredth-place value.** First look at the next place value to the right, (thousandths). It's 6 and it is greater than 5. Thus add 1 to the digit in the hundredth place. It is 3. → $3 + 1 = 4$, then, the answer is 12.84

Round each number to the underlined place value.

1) $32.5\underline{4}8 = 32.55$	2) $2.3\underline{2}6 = 2.33$
3) $55.\underline{4}23 = 55.4$	4) $2\underline{5}.62 = 26$
5) $11.\underline{2}65 = 11.3$	6) $33.5\underline{0}5 = 33.51$
7) $3.5\underline{8}9 = 3.59$	8) $8.0\underline{1}9 = 8.02$

Your Turn!

Name:	Date: ....

Topic	**Adding and Subtracting Decimals**
Notes	✓ Line up the numbers. ✓ Add zeros to have same number of digits for both numbers if necessary. ✓ Add or subtract using column addition or subtraction.
Examples	**Add.** $2.6 + 5.33 =$ First line up the numbers: $\begin{array}{r} 2.6 \\ + 5.33 \end{array}$ →Add zeros to have same number of digits for both numbers. $\begin{array}{r} 2.60 \\ + 5.33 \end{array}$ → Start with the hundredths place. $0 + 3 = 3$, $\begin{array}{r} 2.60 \\ + 5.33 \\ \hline 3 \end{array}$ → Continue with tenths place. $6 + 3 = 9$, $\begin{array}{r} 2.60 \\ + 5.33 \\ \hline .93 \end{array}$ → Add the ones place. $2 + 5 = 7$, $\begin{array}{r} 2.60 \\ + 5.33 \\ \hline 7.93 \end{array}$ **Subtract.** $4.79 - 3.13 =$ $\begin{array}{r} 4.79 \\ - 3.13 \\ \hline \end{array}$ Start with the hundredths place. $9 - 3 = 6$, $\begin{array}{r} 4.79 \\ - 3.13 \\ \hline 6 \end{array}$, continue with tenths place. $7 - 1 = 6$, $\begin{array}{r} 4.79 \\ - 3.13 \\ \hline .66 \end{array}$, subtract the ones place. $4 - 3 = 1$, $\begin{array}{r} 4.79 \\ - 3.13 \\ \hline 1.66 \end{array}$
Your Turn!	1) $48.13 + 20.15 =$　　　　2) $78.14 - 65.19 =$ 3) $38.19 + 24.18 =$　　　　4) $57.26 - 43.54 =$ 5) $27.89 + 46.13 =$　　　　6) $49.65 - 32.78 =$

| Name: .... | Date: .... |

Topic	**Adding and Subtracting Decimals - Answers**
Notes	✓ Line up the numbers. ✓ Add zeros to have same number of digits for both numbers if necessary. ✓ Add or subtract using column addition or subtraction.
Examples	**Add.** $2.6 + 5.33 =$ First line up the numbers: $\begin{array}{r} 2.6 \\ +5.33 \\ \hline \end{array}$ →Add zeros to have same number of digits for both numbers. $\begin{array}{r} 2.60 \\ +5.33 \\ \hline \end{array}$ → Start with the hundredths place. $0 + 3 = 3$, $\begin{array}{r} 2.60 \\ +5.33 \\ \hline 3 \end{array}$ → Continue with tenths place. $6 + 3 = 9$, $\begin{array}{r} 2.60 \\ +5.33 \\ \hline .93 \end{array}$ → Add the ones place. $2 + 5 = 7$, $\begin{array}{r} 2.60 \\ +5.33 \\ \hline 7.93 \end{array}$ **Subtract.** $4.79 - 3.13 =$ $\begin{array}{r} 4.79 \\ -3.13 \\ \hline \end{array}$ Start with the hundredths place. $9 - 3 = 6$, $\begin{array}{r} 4.79 \\ -3.13 \\ \hline 6 \end{array}$, continue with tenths place. $7 - 1 = 6$, $\begin{array}{r} 4.79 \\ -3.13 \\ \hline .66 \end{array}$, subtract the ones place. $4 - 3 = 1$, $\begin{array}{r} 4.79 \\ -3.13 \\ \hline 1.66 \end{array}$
Your Turn!	1) $48.13 + 20.15 = 68.28$ 2) $78.14 - 65.19 = 12.95$ 3) $38.19 + 24.18 = 62.37$ 4) $57.26 - 43.54 = 13.72$ 5) $27.89 + 46.13 = 74.02$ 6) $49.65 - 32.78 = 16.87$

Name: .... Date: ....

Topic	Multiplying and Dividing Decimals
Notes	For Multiplication: ✓ Ignore the decimal point and set up and multiply the numbers as you do with whole numbers. ✓ Count the total number of decimal places in both factors. ✓ Place the decimal point in the product. For Division: ✓ If the divisor is not a whole number, move decimal point to right to make it a whole number. Do the same for dividend. ✓ Divide similar to whole numbers.
Examples	**Find the product.** $1.2 \times 2.3 =$ Set up and multiply the numbers as you do with whole numbers. Line up the numbers: $\frac{12}{\times 23}$ → Multiply: $\frac{\begin{array}{c}12\\ \times 23\end{array}}{276}$ → Count the total number of decimal places in both of the factors. There are two decimal digits. Then: $1.2 \times 2.3 = 2.76$ **Find the quotient.** $5.6 \div 0.8 =$ The divisor is not a whole number. Multiply it by 10 to get 8. → $0.8 \times 10 = 8$ Do the same for the dividend to get 56 → $5.6 \times 10 = 56$ Now, divide: $56 \div 8 = 7$. The answer is 7.
Your Turn!	1) $1.13 \times 0.7 =$ 2) $48.8 \div 8 =$ 3) $0.9 \times 0.68 =$ 4) $66.8 \div 0.2 =$ 5) $0.18 \times 0.5 =$ 6) $37.2 \div 100 =$

Name:	Date: ....

Topic	**Multiplying and Dividing Decimals - Answers**
Notes	For Multiplication: ✓ Ignore the decimal point and set up and multiply the numbers as you do with whole numbers. ✓ Count the total number of decimal places in both factors. ✓ Place the decimal point in the product. For Division: ✓ If the divisor is not a whole number, move decimal point to right to make it a whole number. Do the same for dividend. ✓ Divide similar to whole numbers.
Examples	**Find the product.** $1.2 \times 2.3 =$ Set up and multiply the numbers as you do with whole numbers. Line up the numbers: $\overset{12}{\underset{}{\times 23}}$ → Multiply: $\overset{12}{\underset{276}{\times 23}}$ → Count the total number of decimal places in both of the factors. There are two decimal digits. Then: $1.2 \times 2.3 = 2.76$ **Find the quotient.** $5.6 \div 0.8 =$ The divisor is not a whole number. Multiply it by 10 to get 8. → $0.8 \times 10 = 8$ Do the same for the dividend to get $56 \to 5.6 \times 10 = 56$ Now, divide: $56 \div 8 = 7$. The answer is 7.
Your Turn!	1) $1.13 \times 0.7 = 0.791$ 2) $48.8 \div 8 = 6.1$ 3) $0.9 \times 0.68 = 0.612$ 4) $66.8 \div 0.2 = 334$ 5) $0.18 \times 0.5 = 0.09$ 6) $37.2 \div 100 = 0.372$

Name:	Date: ....

Topic	**Adding and Subtracting Integers**	
Notes	✓ Integers include: zero, counting numbers, and the negative of the counting numbers. $\{..., -3, -2, -1, 0, 1, 2, 3, ...\}$ ✓ Add a positive integer by moving to the right on the number line. ✓ Add a negative integer by moving to the left on the number line. Subtract an integer by adding its opposite.	
Examples	**Solve.** $(4) - (-8) =$ Keep the first number and convert the sign of the second number to its opposite. (change subtraction into addition. Then: $(4) + 8 = 12$ **Solve.** $42 + (12 - 26) =$ First subtract the numbers in brackets, $12 - 26 = -14$ Then: $42 + (-14) = \rightarrow$ change addition into subtraction: $42 - 14 = 28$	
Your Turn!	1) $-(15) + 12 =$	2) $(-2) + (-10) + 18 =$
	3) $(-13) + 7 =$	4) $3 - (-7) + 14 =$
	5) $(-7) + (-8) =$	6) $16 - (-4 + 8) =$
	7) $4 + (-15) + 2 =$	8) $-(22) - (-4) + 8 =$

Name:	Date: ....

Topic	**Adding and Subtracting Integers - Answers**
Notes	✓ Integers include: zero, counting numbers, and the negative of the counting numbers. $\{\ldots, -3, -2, -1, 0, 1, 2, 3, \ldots\}$ ✓ Add a positive integer by moving to the right on the number line. ✓ Add a negative integer by moving to the left on the number line. Subtract an integer by adding its opposite.
Examples	**Solve.** $(4) - (-8) =$ Keep the first number and convert the sign of the second number to its opposite. (change subtraction into addition. Then: $(4) + 8 = 12$ **Solve.** $42 + (12 - 26) =$ First subtract the numbers in brackets, $12 - 26 = -14$ Then: $42 + (-14) = \rightarrow$ change addition into subtraction: $42 - 14 = 28$

Your Turn!	1) $-(15) + 12 = -3$	2) $(-2) + (-10) + 18 = 6$
	3) $(-13) + 7 = -6$	4) $3 - (-7) + 14 = 24$
	5) $(-7) + (-8) = -15$	6) $16 - (-4 + 8) = 12$
	7) $4 + (-15) + 2 = -9$	8) $-(22) - (-4) + 8 = -10$

Name:	Date: ....

Topic	**Multiplying and Dividing Integers**	
Notes	Use following rules for multiplying and dividing integers: ✓ (negative) × (negative) = positive ✓ (negative) ÷ (negative) = positive ✓ (negative) × (positive) = negative ✓ (negative) ÷ (positive) = negative ✓ (positive) × (positive) = positive ✓ (positive) ÷ (negative) = negative	

Examples	**Solve.** $2 \times (14 - 17) =$
	    First subtract the numbers in brackets, $14 - 17 = -3 \rightarrow (2) \times (-3) =$
	    Now use this rule: (positive) × (negative) = negative
	    $(2) \times (-3) = -6$
	
	Solve. $(-7) + (-36 \div 4) =$
	    First divide -36 by 4, the numbers in brackets, using this rule:
	    (negative) ÷ (positive) = negative
	    Then: $-36 \div 4 = -9$. Now, add -7 and -9:
	        $(-7) + (-9) = -7 - 9 = -16$	

Your Turn!	1) $(-7) \times 6 =$	2) $(-63) \div (-7) =$
	3) $(-11) \times (-3) =$	4) $81 \div (-9) =$
	5) $(15 - 12) \times (-7) =$	6) $(-12) \div (3) =$
	7) $4 \times (-9) =$	8) $(8) \div (-2) =$

Name:	Date: ....

Topic	**Multiplying and Dividing Integers - Answers**
Notes	Use following rules for multiplying and dividing integers: ✓ (negative) × (negative) = positive ✓ (negative) ÷ (negative) = positive ✓ (negative) × (positive) = negative ✓ (negative) ÷ (positive) = negative ✓ (positive) × (positive) = positive ✓ (positive) ÷ (negative) = negative
Examples	**Solve.** $2 \times (14 - 17) =$ First subtract the numbers in brackets, $14 - 17 = -3 \rightarrow (2) \times (-3) =$ Now use this rule: (positive) × (negative) = negative $(2) \times (-3) = -6$ **Solve.** $(-7) + (-36 \div 4) =$ First divide -36 by 4, the numbers in brackets, using this rule: (negative) ÷ (positive) = negative Then: $-36 \div 4 = -9$. Now, add -7 and -9: $\qquad (-7) + (-9) = -7 - 9 = -16$
Your Turn!	1) $(-7) \times 6 = -42$ 2) $(-63) \div (-7) = 9$ 3) $(-11) \times (-3) = 33$ 4) $81 \div (-9) = -9$ 5) $(15 - 12) \times (-7) = -21$ 6) $(-12) \div (3) = -4$ 7) $4 \times (-9) = -36$ 8) $(8) \div (-2) = -4$

| Name: .... | Date: .... |

Topic	Order of Operation
Notes	When there is more than one math operation, use PEMDAS: (to memorize this rule, remember the phrase "Please Excuse My Dear Aunt Sally") ✓ Parentheses ✓ Exponents ✓ Multiplication and Division (from left to right) ✓ Addition and Subtraction (from left to right)
Examples	**Calculate.** $(18 - 26) \div (2^4 \div 4) =$ First simplify inside parentheses: $(-8) \div (16 \div 4) = (-8) \div (4)$ Then: $(-8) \div (4) = -2$ **Solve.** $(-5 \times 7) - (18 - 3^2) =$ First calculate within parentheses: $(-5 \times 7) - (18 - 3^2) =$ $(-35) - (18 - 9)$ Then: $(-35) - (18 - 9) = -35 - 9 = -44$
Your Turn!	1) $(11 \times 4) \div (5 + 6) =$ 2) $(30 \div 5) + (17 - 8) =$ 3) $(-9) + (5 \times 6) + 14 =$ 4) $(-10 \times 5) \div (2^2 + 1) =$ 5) $[-16(32 \div 2^3)] \div 8 =$ 6) $(-7) + (72 \div 3^2) + 12 =$ 7) $[16(32 \div 2^3)] - 4^2 =$ 8) $4^3 + (-5 \times 2^5) + 5 =$

Name:	Date: ....

Topic	**Order of Operation - Answers**
Notes	When there is more than one math operation, use PEMDAS: (to memorize this rule, remember the phrase "Please Excuse My Dear Aunt Sally") ✓ Parentheses ✓ Exponents ✓ Multiplication and Division (from left to right) ✓ Addition and Subtraction (from left to right)
Examples	**Calculate.** $(18 - 26) \div (2^4 \div 4) =$ First simplify inside parentheses: $(-8) \div (16 \div 4) = (-8) \div (4)$ Then: $(-8) \div (4) = -2$ **Solve.** $(-5 \times 7) - (18 - 3^2) =$ First calculate within parentheses: $(-5 \times 7) - (18 - 3^2) =$ $(-35) - (18 - 9)$ Then: $(-35) - (18 - 9) = -35 - 9 = -44$

Your Turn!

1) $(11 \times 4) \div (5 + 6) = 4$	2) $(30 \div 5) + (17 - 8) = 15$
3) $(-9) + (5 \times 6) + 14 = 35$	4) $(-10 \times 5) \div (2^2 + 1) = -10$
5) $[-16(32 \div 2^3)] \div 8 = -8$	6) $(-7) + (72 \div 3^2) + 12 = 13$
7) $[16(32 \div 2^3)] - 4^2 = 48$	8) $4^3 + (-5 \times 2^5) + 5 = -91$

Name:	Date: ....

Topic	**Integers and Absolute Value**
Notes	✓ The absolute value of a number is its distance from zero, in either direction, on the number line. For example, the distance of 9 and -9 from zero on number line is 9. ✓ Absolute value is symbolized by vertical bars, as in $|x|$.
Example	**Calculate**. $\mid 8-5 \mid \times \mid 12-16 \mid =$ First calculate $\mid 8-5 \mid$, $\rightarrow \mid 8-5 \mid = \mid 3 \mid$, the absolute value of 3 is 3, $\mid 3 \mid = 3$ $8 \times \mid 12-16 \mid =$ Now calculate $\mid 12-16 \mid$, $\rightarrow \mid 12-16 \mid = \mid -4 \mid$, the absolute value of -4 is 4, $\mid -4 \mid = 4$. Then: $3 \times 4 = 12$

Your Turn!	1) $11 - \mid 4-13 \mid =$	2) $14 - \mid 12-19 \mid - \mid 9 \mid =$
	3) $\mid 21 \mid - \dfrac{\mid -25 \mid}{5} =$	4) $\mid 30 \mid + \dfrac{\mid -49 \mid}{7} =$
	5) $\dfrac{\mid 7 \times -8 \mid}{4} \times \dfrac{\mid -12 \mid}{2} =$	6) $\dfrac{\mid 10 \times -6 \mid}{5} \times \mid -9 \mid =$
	7) $\dfrac{\mid -20 \mid}{5} \times \dfrac{\mid -36 \mid}{6} =$	8) $\mid -30+6 \mid \times \dfrac{\mid -9 \times 4 \mid}{12} =$

Name:	Date: ....

Topic	Integers and Absolute Value - Answers
Notes	✓ The absolute value of a number is its distance from zero, in either direction, on the number line. For example, the distance of 9 and -9 from zero on number line is 9. ✓ Absolute value is symbolized by vertical bars, as in $|x|$.
Example	**Calculate.** $|8 - 5| \times |12 - 16| =$ First calculate $|8 - 5|$, $\rightarrow |8 - 5| = |3|$, the absolute value of 3 is 3, $|3| = 3$ $8 \times |12 - 16| =$ Now calculate $|12 - 16|$, $\rightarrow |12 - 16| = |-4|$, the absolute value of -4 is 4, $|-4| = 4$. Then: $3 \times 4 = 12$
Your Turn!	1) $11 - |4 - 13| = 2$ 2) $14 - |12 - 19| - |9| = -2$ 3) $|21| - \dfrac{|-25|}{5} = 16$ 4) $|30| + \dfrac{|-49|}{7} = 37$ 5) $\dfrac{|7 \times -8|}{4} \times \dfrac{|-12|}{2} = 84$ 6) $\dfrac{|10 \times -6|}{5} \times |-9| = 108$ 7) $\dfrac{|-20|}{5} \times \dfrac{|-36|}{6} = 24$ 8) $|-30 + 6| \times \dfrac{|-9 \times 4|}{12} = 72$

Name:	Date: ....

Topic	**Simplifying Ratios**
Notes	✓ Ratios are used to make comparisons between two numbers. ✓ Ratios can be written as a fraction, using the word "to", or with a colon. ✓ You can calculate equivalent ratios by multiplying or dividing both sides of the ratio by the same number.
Examples	**Simplify. $18:63 =$** Both numbers 18 and 63 are divisible by $9 \Rightarrow 18 \div 9 = 2, 63 \div 9 = 7$, Then: $18:63 = 2:7$ **Simplify. $\frac{25}{45} =$** Both numbers 25 and 45 are divisible by $5, \Rightarrow 25 \div 5 = 5, 45 \div 5 = 9$, Then: $\frac{25}{45} = \frac{5}{9}$

Your Turn!	1) $\frac{4}{32} = -$	2) $\frac{25}{80} = -$
	3) $\frac{15}{35} = -$	4) $\frac{42}{54} = -$
	5) $\frac{12}{36} = -$	6) $\frac{30}{80} = -$
	7) $\frac{18}{24} = -$	8) $\frac{60}{108} = -$

Name:	Date: ....

Topic	**Simplifying Ratios - Answers**
Notes	✓ Ratios are used to make comparisons between two numbers. ✓ Ratios can be written as a fraction, using the word "to", or with a colon. ✓ You can calculate equivalent ratios by multiplying or dividing both sides of the ratio by the same number.
Examples	**Simplify. $18:63 =$** Both numbers 18 and 63 are divisible by $9 \Rightarrow 18 \div 9 = 2, 63 \div 9 = 7$, Then: $18:63 = 2:7$ **Simplify. $\dfrac{25}{45} =$** Both numbers 25 and 45 are divisible by 5, $\Rightarrow 25 \div 5 = 5, 45 \div 5 = 9$, Then: $\dfrac{25}{45} = \dfrac{5}{9}$
Your Turn!	1) $\dfrac{4}{32} = \dfrac{1}{8}$ 2) $\dfrac{25}{80} = \dfrac{5}{16}$ 3) $\dfrac{15}{35} = \dfrac{3}{7}$ 4) $\dfrac{42}{54} = \dfrac{7}{9}$ 5) $\dfrac{12}{36} = \dfrac{1}{3}$ 6) $\dfrac{30}{80} = \dfrac{3}{8}$ 7) $\dfrac{18}{24} = \dfrac{3}{4}$ 8) $\dfrac{60}{108} = \dfrac{5}{9}$

Name:	Date: ....

Topic	Proportional Ratios
Notes	✓ Two ratios are proportional if they represent the same relationship. ✓ A proportion means that two ratios are equal. It can be written in two ways: $\dfrac{a}{b} = \dfrac{c}{d}$ $\qquad a : b = c : d$
Example	**Solve this proportion for** x. $\dfrac{5}{8} = \dfrac{35}{x}$ Use cross multiplication: $\dfrac{5}{8} = \dfrac{35}{x} \Rightarrow 5 \times x = 8 \times 35 \Rightarrow 5x = 280$ Divide to find x: $\quad x = \dfrac{280}{5} \Rightarrow x = 56$
Your Turn!	1) $\dfrac{1}{9} = \dfrac{8}{x} \Rightarrow x = $ _____ 2) $\dfrac{5}{8} = \dfrac{25}{x} \Rightarrow x = $ _____ 3) $\dfrac{3}{11} = \dfrac{6}{x} \Rightarrow x = $ _____ 4) $\dfrac{12}{20} = \dfrac{x}{200} \Rightarrow x = $ _____ 5) $\dfrac{9}{12} = \dfrac{27}{x} \Rightarrow x = $ _____ 6) $\dfrac{14}{16} = \dfrac{x}{80} \Rightarrow x = $ _____ 7) $\dfrac{7}{15} = \dfrac{49}{x} \Rightarrow x = $ _____ 8) $\dfrac{8}{19} = \dfrac{32}{x} \Rightarrow x = $ _____

Name:	Date: ....

Topic	**Proportional Ratios - Answers**
Notes	✓ Two ratios are proportional if they represent the same relationship. ✓ A proportion means that two ratios are equal. It can be written in two ways: $\dfrac{a}{b} = \dfrac{c}{d}$ $\qquad a : b = c : d$
Example	**Solve this proportion for** x. $\dfrac{5}{8} = \dfrac{35}{x}$ Use cross multiplication: $\dfrac{5}{8} = \dfrac{35}{x} \Rightarrow 5 \times x = 8 \times 35 \Rightarrow 5x = 280$ Divide to find x: $x = \dfrac{280}{5} \Rightarrow x = 56$

Your Turn!	1) $\dfrac{1}{9} = \dfrac{8}{x} \Rightarrow x = 72$	2) $\dfrac{5}{8} = \dfrac{25}{x} \Rightarrow x = 40$
	3) $\dfrac{3}{11} = \dfrac{6}{x} \Rightarrow x = 22$	4) $\dfrac{12}{20} = \dfrac{x}{200} \Rightarrow x = 120$
	5) $\dfrac{9}{12} = \dfrac{27}{x} \Rightarrow x = 36$	6) $\dfrac{14}{16} = \dfrac{x}{80} \Rightarrow x = 70$
	7) $\dfrac{7}{15} = \dfrac{49}{x} \Rightarrow x = 105$	8) $\dfrac{8}{19} = \dfrac{32}{x} \Rightarrow x = 76$

Name:	Date: ....

Topic	Create Proportion
Notes	✓ To create a proportion, simply find (or create) two equal fractions. ✓ Use cross products to solve proportions or to test whether two ratios are equal and form a proportion. $\frac{a}{b} = \frac{c}{d} \Rightarrow a \times d = c \times b$
Example	**State if this pair of ratios form a proportion. $\frac{2}{3}$ and $\frac{12}{30}$** Use cross multiplication: $\frac{2}{3} = \frac{12}{30} \rightarrow 2 \times 30 = 12 \times 3 \rightarrow 60 = 36$, which is not correct. Therefore, this pair of ratios doesn't form a proportion.
Your Turn!	**State if each pair of ratios form a proportion.** 1) $\frac{4}{8}$ and $\frac{24}{48}$ 2) $\frac{5}{15}$ and $\frac{10}{20}$ 3) $\frac{3}{11}$ and $\frac{9}{33}$ 4) $\frac{7}{10}$ and $\frac{14}{20}$ 5) $\frac{7}{9}$ and $\frac{48}{81}$ 6) $\frac{6}{8}$ and $\frac{12}{14}$ 7) $\frac{2}{10}$ and $\frac{6}{30}$ 8) $\frac{9}{12}$ and $\frac{18}{24}$ 9) Solve. Five pencils costs \$0.65. How many pencils can you buy for \$2.60? _____

Name:	Date: ....

Topic	**Create Proportion - Answers**
Notes	✓ To create a proportion, simply find (or create) two equal fractions. ✓ Use cross products to solve proportions or to test whether two ratios are equal and form a proportion. $\frac{a}{b} = \frac{c}{d} \Rightarrow a \times d = c \times b$
Example	**State if this pair of ratios form a proportion. $\frac{2}{3}$ and $\frac{12}{30}$** Use cross multiplication: $\frac{2}{3} = \frac{12}{30} \rightarrow 2 \times 30 = 12 \times 3 \rightarrow 60 = 36$, which is not correct. Therefore, this pair of ratios doesn't form a proportion.
Your Turn!	**State if each pair of ratios form a proportion.** 1) $\frac{4}{8}$ and $\frac{24}{48}$, *Yes* 2) $\frac{5}{15}$ and $\frac{10}{20}$, *No* 3) $\frac{3}{11}$ and $\frac{9}{33}$, *Yes* 4) $\frac{7}{10}$ and $\frac{14}{20}$, *Yes* 5) $\frac{7}{9}$ and $\frac{48}{81}$, *No* 6) $\frac{6}{8}$ and $\frac{12}{14}$, *No* 7) $\frac{2}{10}$ and $\frac{6}{30}$, *Yes* 8) $\frac{9}{12}$ and $\frac{18}{24}$, *Yes* 9) Solve. Five pencils costs \$0.65. How many pencils can you buy for \$2.60? *20 pencils*

Name:	Date: ....

Topic	Similarity and Ratios
Notes	✓ Two figures are similar if they have the same shape. ✓ Two or more figures are similar if the corresponding angles are equal, and the corresponding sides are in proportion.
Example	**Following triangles are similar. What is the value of unknown side?** **Solution:** Find the corresponding sides and write a proportion: $\frac{4}{12} = \frac{x}{9}$. Now, use cross product to solve for x: $\frac{4}{12} = \frac{x}{9} \rightarrow 4 \times 9 = 12 \times x \rightarrow$ $36 = 12x$. Divide both sides by 12. Then: $36 = 12x \rightarrow \frac{36}{12} = \frac{12x}{12} \rightarrow x = 3$. The missing side is 3.

Your Turn!

1)

2)

3)

4)

5)

6)

| Name: .... | Date: .... |

Topic	**Similarity and Ratios - Answers**
Notes	✓ Two figures are similar if they have the same shape. ✓ Two or more figures are similar if the corresponding angles are equal, and the corresponding sides are in proportion.
Example	**Following triangles are similar. What is the value of unknown side?** **Solution:** Find the corresponding sides and write a proportion: $\frac{4}{12} = \frac{x}{9}$. Now, use cross product to solve for x: $\frac{4}{12} = \frac{x}{9} \rightarrow 4 \times 9 = 12 \times x \rightarrow$ $36 = 12x$. Divide both sides by 12. Then: $36 = 12x \rightarrow \frac{36}{12} = \frac{12x}{12} \rightarrow x = 3$. The missing side is 3.

| **Your Turn!** | 1) 24

 2) 11

 3) 4

 4) 8

 5) 10

 6) 9 |

Name:	Date: ....

Topic	**Percent Problems**
Notes	✓ In each percent problem, we are looking for the base, or part or the percent. ✓ Use the following equations to find each missing section. ○ Base = Part ÷ Percent ○ Part = Percent × Base ○ Percent = Part ÷ Base
Examples	**18 is what percent of 30?** In this problem, we are looking for the percent. Use the following equation: $Percent = Part ÷ Base → Percent = 18 ÷ 30 = 0.6 = 60\%$ **40 is 20% of what number?** Use the following formula: $Base = Part ÷ Percent →$ $Base = 40 ÷ 0.20 = 200$ 40 is 20% of 200.

Your Turn!	1) What is 25 percent of 800?	2) 26 is what percent of 200?
	3) 60 is 5 percent of what number?	4) 48 is what percent of 300?
	5) 84 is 28 percent of what number?	6) 63 is what percent of 700?
	7) 96 is 24 percent of what number?	8) 40 is what percent of 800?

| Name: .... | Date: .... |

Topic	**Percent Problems – Answers**

Notes	✓ In each percent problem, we are looking for the base, or part or the percent.
	✓ Use the following equations to find each missing section.
	    ○ Base = Part ÷ Percent
	    ○ Part = Percent × Base
	    ○ Percent = Part ÷ Base

Examples	**18 is what percent of 30?**
	    In this problem, we are looking for the percent. Use the following equation:
	$Percent = Part \div Base \rightarrow Percent = 18 \div 30 = 0.6 = 60\%$
	
	40 is 20% of what number?
	    Use the following formula: $Base = Part \div Percent \rightarrow$
	$Base = 40 \div 0.20 = 200$
	40 is 20% of 200.

Your Turn!	1) What is 25 percent of 800? 200 2) 26 is what percent of 200? 13% 3) 60 is 5 percent of what number? 1,200 4) 48 is what percent of 300? 16% 5) 84 is 28 percent of what number? 300 6) 63 is what percent of 700? 9% 7) 96 is 24 percent of what number? 400 8) 40 is what percent of 800? 5%

Name:	Date: ....

Topic	**Percent of Increase and Decrease**
Notes	✓ Percent of change (increase or decrease) is a mathematical concept that represents the degree of change over time. ✓ To find the percentage of increase or decrease: 1- New Number – Original Number 2- The result ÷ Original Number × 100 Or use this formula: $Percent\ of\ change = \frac{new\ number - original\ number}{original\ number} \times 100$
Example	**The price of a printer increases from $40 to $50. What is the percent increase?** **Solution:** $Percent\ of\ change = \frac{new\ number - original\ number}{original\ number} \times 100 =$ $\frac{50-40}{40} \times 100 = 25$ The percentage increase is 25. It means that the price of the printer increased 25%.
Your Turn!	1) In a class, the number of students has been increased from 32 to 36. What is the percentage increase? _____ % 2) The price of gasoline rose from $4.50 to $5.40 in one month. By what percent did the gas price rise? _____ % 3) A shirt was originally priced at $65.00. It went on sale for $52.00. What was the percent that the shirt was discounted? _____ % 4) Jason got a raise, and his hourly wage increased from $40 to $52. What is the percent increase? _____ %

Name:	Date: ....

Topic	Percent of Increase and Decrease - Answers

Notes	✓ Percent of change (increase or decrease) is a mathematical concept that represents the degree of change over time.
	
	✓ To find the percentage of increase or decrease:
	
	  1- New Number − Original Number
	
	  2- The result ÷ Original Number × 100
	
	Or use this formula: $Percent\ of\ change = \dfrac{new\ number\ -\ original\ number}{original\ number} \times 100$

Example	**The price of a printer increases from \$40 to \$50. What is the percent increase?** **Solution:** $Percent\ of\ change = \dfrac{new\ number\ -\ original\ number}{original\ number} \times 100 =$ $\dfrac{50-40}{40} \times 100 = 25$ The percentage increase is 25. It means that the price of the printer increased 25%.
Your Turn!	1) In a class, the number of students has been increased from 32 to 36. What is the percentage increase? 12.5% 2) The price of gasoline rose from \$4.50 to \$5.40 in one month. By what percent did the gas price rise? 20% 3) A shirt was originally priced at \$65.00. It went on sale for \$52.00. What was the percent that the shirt was discounted? 20% 4) Jason got a raise, and his hourly wage increased from \$40 to \$52. What is the percent increase? 30%

| Name: .... | Date: .... |

Topic	Discount, Tax and Tip
Notes	✓ Discount = Multiply the regular price by the rate of discount ✓ Selling price = original price − discount ✓ To find tax, multiply the tax rate to the taxable amount (income, property value, etc.) ✓ To find tip, multiply the rate to the selling price.
Example	**The original price of a table is $300 and the tax rate is 6%. What is the final price of the table?** **Solution:** First find the tax amount. To find tax: Multiply the tax rate to the taxable amount. Tax rate is 6% or 0.06. Then: $0.06 \times 300 = 18$. The tax amount is $18. Final price is: $300 + $18 = $318

Your Turn!	1) Original price of a chair: $300 Tax: 15%, Selling price: _____	2) Original price of a computer: $750 Discount: 20%, Selling price: _____
	3) Original price of a printer: $250 Tax: 10%, Selling price: _____	4) Original price of a sofa: $620 Discount: 25%, Selling price: _____
	5) Original price of a mattress: $800 Tax: 12%, Selling price: _____	6) Original price of a book: $150 Discount: 60%, Selling price: _____
	7) Restaurant bill: $35.00 Tip: 20%, Final amount: _____	8) Restaurant bill: $60.00 Tip: 25%, Final amount: _____

Name:	Date: ....

Topic	Discount, Tax and Tip - Answers
Notes	✓ Discount = Multiply the regular price by the rate of discount ✓ Selling price = original price – discount ✓ To find tax, multiply the tax rate to the taxable amount (income, property value, etc.) ✓ To find tip, multiply the rate to the selling price.
Example	**The original price of a table is $300 and the tax rate is 6%. What is the final price of the table?** **Solution:** First find the tax amount. To find tax: Multiply the tax rate to the taxable amount. Tax rate is 6% or 0.06. Then: $0.06 \times 300 = 18$. The tax amount is $18. Final price is: $300 + $18 = $318

Your Turn!		
	1) Original price of a chair: $300 Tax: 15%, Selling price: $345	2) Original price of a computer: $750 Discount: 20%, Selling price: $600
	3) Original price of a printer: $250 Tax: 10%, Selling price: $275	4) Original price of a sofa: $620 Discount: 25%, Selling price: $465
	5) Original price of a mattress: $800 Tax: 12%, Selling price: $896	6) Original price of a book: $150 Discount: 60%, Selling price: $60
	7) Restaurant bill: $35.00 Tip: 20%, Final amount: $42	8) Restaurant bill: $60.00 Tip: 25%, Final amount: $75

Name:	Date: ....

Topic	Simple Interest
Notes	✓ Simple Interest: The charge for borrowing money or the return for lending it. To solve a simple interest problem, use this formula: Interest = principal x rate x time ⇒ $I = p \times r \times t$
Example	**Find simple interest for $3,000 investment at 5% for 4 years.** **Solution:** Use Interest formula: $I = prt$ ($p = \$3,000$, $r = 5\% = 0.05$ and $t = 4$) Then: $I = 3,000 \times 0.05 \times 4 = \$600$

Your Turn!	1) $250 at 4% for 3 years. Simple interest: $_____	2) $3,300 at 5% for 6 years. Simple interest: $_____
	3) $720 at 2% for 5 years. Simple interest: $_____	4) $2,200 at 8% for 4 years. Simple interest: $_____
	5) $1,800 at 3% for 2 years. Simple interest: $_____	6) $530 at 4% for 5 years. Simple interest: $_____
	7) $7,000 at 5% for 3 months. Simple interest: $_____	8) $880 at 5% for 9 months. Simple interest: $_____

| Name: .... | Date: .... |

Topic	Simple Interest - Answers
Notes	✓ Simple Interest: The charge for borrowing money or the return for lending it. To solve a simple interest problem, use this formula: Interest = principal x rate x time ⇒ $I = p \times r \times t$
Example	**Find simple interest for $3,000 investment at 5% for 4 years.** **Solution:** Use Interest formula: $I = prt$ ($p = \$3,000$, $r = 5\% = 0.05$ and $t = 4$) Then: $I = 3,000 \times 0.05 \times 4 = \$600$

Your Turn!

1) $250 at 4% for 3 years.
   Simple interest: $30

2) $3,300 at 5% for 6 years.
   Simple interest: $990

3) $720 at 2% for 5 years.
   Simple interest: $72

4) $2,200 at 8% for 4 years.
   Simple interest: $704

5) $1,800 at 3% for 2 years.
   Simple interest: $108

6) $530 at 4% for 5 years.
   Simple interest: $106

7) $7,000 at 5% for 3 months.
   Simple interest: $87.50

8) $880 at 5% for 9 months.
   Simple interest: $33

| Name: .... | Date: .... |

Topic	Simplifying Variable Expressions
Notes	✓ In algebra, a variable is a letter used to stand for a number. The most common letters are: $x, y, z, a, b, c, m,$ and n. ✓ Algebraic expression is an expression contains integers, variables, and the math operations such as addition, subtraction, multiplication, division, etc. ✓ In an expression, we can combine "like" terms. (values with same variable and same power)
Example	**Simplify this expression.** $(6x + 8x + 9) = ?$ Combine like terms. Then: $(6x + 8x + 9) = 14x + 9$ (remember you cannot combine variables and numbers).

Your Turn!	1) $5x + 2 - 2x =$	2) $4 + 7x + 3x =$
	3) $8x + 3 - 3x =$	4) $-2 - x^2 - 6x^2 =$
	5) $3 + 10x^2 + 2 =$	6) $8x^2 + 6x + 7x^2 =$
	7) $5x^2 - 12x^2 + 8x =$	8) $2x^2 - 2x - x + 5x^2 =$
	9) $4x - (12 - 30x) =$	10) $10x - (80x - 48) =$

Name:	Date: ....

Topic	**Simplifying Variable Expressions - Answers**
Notes	✓ In algebra, a variable is a letter used to stand for a number. The most common letters are: x, y, z, a, b, c, m, and n. ✓ Algebraic expression is an expression contains integers, variables, and the math operations such as addition, subtraction, multiplication, division, etc. ✓ In an expression, we can combine "like" terms. (values with same variable and same power)
Example	**Simplify this expression.** $(6x + 8x + 9) = ?$ Combine like terms. Then: $(6x + 8x + 9) = 14x + 9$ (remember you cannot combine variables and numbers).

Your Turn!	1) $5x + 2 - 2x =$ $3x + 2$	2) $4 + 7x + 3x =$ $10x + 4$
	3) $8x + 3 - 3x =$ $5x + 3$	4) $-2 - x^2 - 6x^2 =$ $-7x^2 - 2$
	5) $3 + 10x^2 + 2 =$ $10x^2 + 5$	6) $8x^2 + 6x + 7x^2 =$ $15x^2 + 6x$
	7) $5x^2 - 12x^2 + 8x =$ $-7x^2 + 8x$	8) $2x^2 - 2x - x + 5x^2 =$ $7x^2 - 3x$
	9) $4x - (12 - 30x) =$ $34x - 12$	10) $10x - (80x - 48) =$ $-70x + 48$

Name:	Date: ....

Topic	**Simplifying Polynomial Expressions**	
Notes	✓ In mathematics, a polynomial is an expression consisting of variables and coefficients that involves only the operations of addition, subtraction, multiplication, and non–negative integer exponents of variables. $$P(x) = a_n x^n + a_{n-1} x^{n-1} + \ldots + a_2 x^2 + a_1 x + a_0$$	
Example	**Simplify this expression.** $(2x^2 - x^4) - (4x^4 - x^2) =$ First use distributive property:　→ multiply $(-)$ into $(4x^4 - x^2)$ $(2x^2 - x^4) - (4x^4 - x^2) = 2x^2 - x^4 - 4x^4 + x^2$ Then combine "like" terms: $2x^2 - x^4 - 4x^4 + x^2 = 3x^2 - 5x^4$ And write in standard form: $3x^2 - 5x^4 = -5x^4 + 3x^2$	
Your Turn!	1) $(2x^3 + 5x^2) - (12x + 2x^2) =$	2) $(2x^5 + 2x^3) - (7x^3 + 6x^2) =$
	3) $(12x^4 + 4x^2) - (2x^2 - 6x^4) =$	4) $14x - 3x^2 - 2(6x^2 + 6x^3) =$
	5) $(5x^3 - 3) + 5(2x^2 - 3x^3) =$	6) $(4x^3 - 2x) - 2(4x^3 - 2x^4) =$
	7) $2(4x - 3x^3) - 3(3x^3 + 4x^2) =$	8) $(2x^2 - 2x) - (2x^3 + 5x^2) =$

Name:	Date: ....

Topic	Simplifying Polynomial Expressions - Answers
Notes	✓ In mathematics, a polynomial is an expression consisting of variables and coefficients that involves only the operations of addition, subtraction, multiplication, and non–negative integer exponents of variables. $$P(x) = a_{\mathrm{n}}x^n + a_{\mathrm{n-1}}x^{n-1} + \dots + a_2x^2 + a_1x + a_0$$
Example	**Simplify this expression.** $(2x^2 - x^4) - (4x^4 - x^2) =$ First use distributive property: → multiply $(-)$ into $(4x^4 - x^2)$ $(2x^2 - x^4) - (4x^4 - x^2) = 2x^2 - x^4 - 4x^4 + x^2$ Then combine "like" terms: $2x^2 - x^4 - 4x^4 + x^2 = 3x^2 - 5x^4$ And write in standard form: $3x^2 - 5x^4 = -5x^4 + 3x^2$

Your Turn!	1) $(2x^3 + 5x^2) - (12x + 2x^2) =$ $2x^3 + 3x^2 - 12x$	2) $(2x^5 + 2x^3) - (7x^3 + 6x^2) =$ $2x^5 - 5x^3 - 6x^2$
	3) $(12x^4 + 4x^2) - (2x^2 - 6x^4) =$ $18x^4 + 2x^2$	4) $14x - 3x^2 - 2(6x^2 + 6x^3) =$ $-12x^3 - 15x^2 + 14x$
	5) $(5x^3 - 3) + 5(2x^2 - 3x^3) =$ $-10x^3 + 10x^2 - 3$	6) $(4x^3 - 2x) - 2(4x^3 - 2x^4) =$ $4x^4 - 4x^3 - 2$
	7) $2(4x - 3x^3) - 3(3x^3 + 4x^2) =$ $-15x^3 - 12x^2 + 8x$	8) $(2x^2 - 2x) - (2x^3 + 5x^2) =$ $-2x^3 - 3x^2 - 2x$

| Name: .... | Date: .... |

Topic	**Evaluating One Variable**
Notes	✓ To evaluate one variable expression, find the variable and substitute a number for that variable. ✓ Perform the arithmetic operations.
Example	**Find the value of this expression for $x = -3$. $-3x - 13$** **Solution:** Substitute -3 for x, then: $-3x - 13 = -3(-3) - 13 = 9 - 13 = -4$

Your Turn!	1) $x = -3 \Rightarrow 3x + 8 = $ _____	2) $x = 4 \Rightarrow 4(2x + 6) = $ _____
	3) $x = -1 \Rightarrow 6x + 4 = $ _____	4) $x = 7 \Rightarrow 6(5x + 3) = $ _____
	5) $x = 4 \Rightarrow 5(3x + 2) = $ ___	6) $x = 6 \Rightarrow 3(2x + 4) = $ _____
	7) $x = 3 \Rightarrow 7(3x + 1) = $ ___	8) $x = 8 \Rightarrow 3(3x + 7) = $ _____
	9) $x = 9 \Rightarrow 2(x + 9) = $ _____	10) $x = 7 \Rightarrow 2(4x + 5) = $ _____

Name:	Date: ....

Topic	**Evaluating One Variable - Answers**	
Notes	✓ To evaluate one variable expression, find the variable and substitute a number for that variable. ✓ Perform the arithmetic operations.	
Example	**Find the value of this expression for** $x = -3. -3x - 13$ **Solution:** Substitute -3 for x, then: $-3x - 13 = -3(-3) - 13 = 9 - 13 = -4$	
Your Turn!	1) $x = -3 \Rightarrow 3x + 8 = -1$	2) $x = 4 \Rightarrow 4(2x + 6) = 56$
	3) $x = -1 \Rightarrow 6x + 4 = -2$	4) $x = 7 \Rightarrow 6(5x + 3) = 228$
	5) $x = 4 \Rightarrow 5(3x + 2) = 70$	6) $x = 6 \Rightarrow 3(2x + 4) = 48$
	7) $x = 3 \Rightarrow 7(3x + 1) = 70$	8) $x = 8 \Rightarrow 3(3x + 7) = 93$
	9) $x = 9 \Rightarrow 2(x + 9) = 36$	10) $x = 7 \Rightarrow 2(4x + 5) = 66$

Name:	Date: ....

Topic	**Evaluating Two Variables**	
Notes	✓ To evaluate an algebraic expression, substitute a number for each variable. ✓ Perform the arithmetic operations to find the value of the expression.	
Example	**Evaluate this expression for $a = 4$ and $b = -2$. $5a - 6b$** **Solution:** Substitute 4 for a, and -2 for b , then: $5a - 6b = 5(4) - 6(-2) = 20 + 12 = 32$	
Your Turn!	1) $-4a + 6b$, $a = 4$, $b = 3$ _____	2) $5x + 3y$, $x = 2$, $y = -1$ _____
	3) $-5a + 3b$, $a = 2$, $b = -2$ _____	4) $3x - 4y$, $x = 6$, $y = 2$ _____
	5) $2z + 14 + 6k$, $z = 5$, $k = 3$ _____	6) $7a - (9 - 3b)$, $a = 1$, $b = 1$ _____
	7) $-6a + 3b$, $a = 4$, $b = 3$ _____	8) $-2a + b$, $a = 6$, $b = 9$ _____
	9) $8x + 2y$, $x = 4$, $y = 5$ _____	10) $z + 4 + 2k$, $z = 7$, $k = 4$ _____

Name:	Date: ....

Topic	**Evaluating Two Variables - Answers**
Notes	✓ To evaluate an algebraic expression, substitute a number for each variable. ✓ Perform the arithmetic operations to find the value of the expression.
Example	**Evaluate this expression for $a = 4$ and $b = -2$. $5a - 6b$** **Solution:** Substitute 4 for a, and -2 for b , then: $5a - 6b = 5(4) - 6(-2) = 20 + 12 = 32$

Your Turn!	1) $-4a + 6b$, $a = 4$, $b = 3$ 2	2) $5x + 3y$, $x = 2$, $y = -1$ 7
	3) $-5a + 3b$, $a = 2$, $b = -2$ -16	4) $3x - 4y$, $x = 6$, $y = 2$ 10
	5) $2z + 14 + 6k$, $z = 5$, $\qquad\qquad\qquad k = 3$ 42	6) $7a - (9 - 3b)$, $a = 1$, $\qquad\qquad\qquad b = 1$ 1
	7) $-6a + 3b$, $a = 4$, $b = 3$ -15	8) $-2a + b$, $a = 6$, $b = 9$ -3
	9) $8x + 2y$, $x = 4$, $y = 5$ 42	10) $z + 4 + 2k$, $z = 7$, $k = 4$ 19

Name:	Date: ....

Topic	**The Distributive Property**
Notes	✓ The distributive property (or the distributive property of multiplication over addition and subtraction) simplifies and solves expressions in the form of: $a(b + c)$ or $a(b - c)$ ✓ Distributive Property rule: $$a(b + c) = ab + ac$$
Example	**Simply.** $(5)(2x - 8)$ **Solution:** Use Distributive Property rule: $a(b + c) = ab + ac$ $$(5)(2x - 8) = (5 \times 2x) + (5) \times (-8) = 10x - 40$$

Your Turn!	1) $(-2)(4 - 3x) =$	2) $(6 - 3x)(-7)$
	3) $6(5 - 9x) =$	4) $10(3 - 5x) =$
	5) $5(6 - 5x) =$	6) $(-2)(-5x + 3) =$
	7) $(8 - 9x)(5) =$	8) $(-16x + 15)(-3) =$
	9) $(-2x + 7)(3) =$	10) $(-18x + 25)(-2) =$

Name:	Date: ....

Topic	The Distributive Property - Answers
Notes	✓ The distributive property (or the distributive property of multiplication over addition and subtraction) simplifies and solves expressions in the form of: $a(b + c)$ or $a(b - c)$ ✓ Distributive Property rule: $$a(b + c) = ab + ac$$
Example	**Simply.** $(5)(2x - 8)$ **Solution:** Use Distributive Property rule: $a(b + c) = ab + ac$ $$(5)(2x - 8) = (5 \times 2x) + (5) \times (-8) = 10x - 40$$
Your Turn!	1) $(-2)(4 - 3x) = 6x - 8$ 2) $(6 - 3x)(-7) = 21x - 42$ 3) $6(5 - 9x) = -54x + 30$ 4) $10(3 - 5x) = -50x + 30$ 5) $5(6 - 5x) = -25x + 30$ 6) $(-2)(-5x + 3) = 10x - 6$ 7) $(8 - 9x)(5) = -45x + 40$ 8) $(-16x + 15)(-3) = 48x - 45$ 9) $(-2x + 7)(3) = -6x + 21$ 10) $(-18x + 25)(-2) = 36x - 50$

Name:	Date: ....

Topic	One–Step Equations

Notes	✓ You only need to perform one Math operation in order to solve the one-step equations.
	✓ To solve one-step equation, find the inverse (opposite) operation is being performed.
	✓ The inverse operations are:
	   - Addition and subtraction
	   - Multiplication and division

Example	**Solve this equation.** $x + 42 = 60 \Rightarrow x = ?$ Here, the operation is addition and its inverse operation is subtraction. To solve this equation, subtract 42 from both sides of the *equation:* $x + 42 - 42 = 60 - 42$ Then simplify: $x + 42 - 42 = 60 - 42 \Rightarrow x = 18$
Your Turn!	1) $x - 15 = 36 \Rightarrow x = $ ____ 2) $18 = 13 + x \Rightarrow x = $ ____ 3) $x - 22 = 54 \Rightarrow x = $ ____ 4) $x + 14 = 24 \Rightarrow x = $ ____ 5) $4x = 24 \Rightarrow x = $ ____ 6) $\frac{x}{6} = -3 \Rightarrow x = $ ____ 7) $99 = 11x \Rightarrow x = $ ____ 8) $\frac{x}{12} = 6 \Rightarrow x = $ ____

Name:	Date: ....

Topic	One–Step Equations - Answers

Notes	✓ You only need to perform one Math operation in order to solve the one-step equations.
	✓ To solve one-step equation, find the inverse (opposite) operation is being performed.
	✓ The inverse operations are:
	  - Addition and subtraction
	  - Multiplication and division

Example	**Solve this equation.** $x + 42 = 60 \Rightarrow x = ?$ Here, the operation is addition and its inverse operation is subtraction. To solve this equation, subtract 42 from both sides of the *equation:* $x + 42 - 42 = 60 - 42$ Then simplify: $x + 42 - 42 = 60 - 42 \Rightarrow x = 18$

Your Turn!	1) $x - 15 = 36 \Rightarrow x = 51$	2) $18 = 13 + x \Rightarrow x = 5$
	3) $x - 22 = 54 \Rightarrow x = 76$	4) $x + 14 = 24 \Rightarrow x = 10$
	5) $4x = 24 \Rightarrow x = 6$	6) $\frac{x}{6} = -3 \Rightarrow x = -18$
	7) $99 = 11x \Rightarrow x = 9$	8) $\frac{x}{12} = 6 \Rightarrow x = 72$

Name:	Date: ....

Topic	**Multi –Step Equations**
Notes	✓ Combine "like" terms on one side. ✓ Bring variables to one side by adding or subtracting. ✓ Simplify using the inverse of addition or subtraction. ✓ Simplify further by using the inverse of multiplication or division. ✓ Check your solution by plugging the value of the variable into the original equation.
Example	**Solve this equation for** x. $2x - 3 = 13$ **Solution:** The inverse of subtraction is addition. Add 3 to both sides of the equation. Then: $2x - 3 = 13 \Rightarrow 2x - 3 + 3 = 13 + 3 \Rightarrow 2x = 16$ Now, divide both sides by 2, then: $\frac{2x}{2} = \frac{16}{2} \Rightarrow x = 8$ Now, check the solution: $x = 8 \Rightarrow 2x - 3 = 13 \Rightarrow 2(8) - 3 = 13 \Rightarrow$ $16 - 3 = 13$ The answer $x = 8$ is correct.

Your Turn!	1) $4x - 12 = 8 \Rightarrow x =$	2) $12 - 3x = -6 + 3x \Rightarrow x =$
	3) $3(4 - 2x) = 24 \Rightarrow x =$	4) $15 + 5x = -7 - 6x \Rightarrow x =$
	5) $-2(5 + x) = 2 \Rightarrow x =$	6) $12 - 2x = -3 - 5x \Rightarrow x =$
	7) $14 = -(x - 9) \Rightarrow x =$	8) $11 - 4x = -4 - 3x \Rightarrow x =$

Name:	Date: ....

Topic	**Multi –Step Equations - Answers**
Notes	✓ Combine "like" terms on one side. ✓ Bring variables to one side by adding or subtracting. ✓ Simplify using the inverse of addition or subtraction. ✓ Simplify further by using the inverse of multiplication or division. ✓ Check your solution by plugging the value of the variable into the original equation.
Example	**Solve this equation for** x. $2x - 3 = 13$ **Solution:** The inverse of subtraction is addition. Add 3 to both sides of the equation. Then: $2x - 3 = 13 \Rightarrow 2x - 3 + 3 = 13 + 3 \Rightarrow 2x = 16$ Now, divide both sides by 2, then: $\frac{2x}{2} = \frac{16}{2} \Rightarrow x = 8$ Now, check the solution: $x = 8 \Rightarrow 2x - 3 = 13 \Rightarrow 2(8) - 3 = 13 \Rightarrow$ $16 - 3 = 13$ The answer $x = 8$ is correct.

Your Turn!	1) $4x - 12 = 8 \Rightarrow x = 5$	2) $12 - 3x = -6 + 3x \Rightarrow x = 3$
	3) $3(4 - 2x) = 24 \Rightarrow x = -2$	4) $15 + 5x = -7 - 6x \Rightarrow x = -2$
	5) $-2(5 + x) = 2 \Rightarrow x = -6$	6) $12 - 2x = -3 - 5x \Rightarrow x = -5$
	7) $14 = -(x - 9) \Rightarrow x = -5$	8) $11 - 4x = -4 - 3x \Rightarrow x = 15$

Name:	Date: ....

Topic	System of Equations
Notes	✓ A system of equations contains two equations and two variables. For example, consider the system of equations: $x - 2y = -2, x + 2y = 10$ ✓ The easiest way to solve a system of equation is using the elimination method. The elimination method uses the addition property of equality. You can add the same value to each side of an equation. ✓ For the first equation above, you can add $x + 2y$ to the left side and 10 to the right side of the first equation: $x - 2y + (x + 2y) = -2 + 10$. Now, if you simplify, you get: $x - 2y + (x + 2y) = -2 + 10 \rightarrow 2x = 8 \rightarrow x = 4$. Now, substitute 4 for the x in the first equation: $4 - 2y = -2$. By solving this equation, $y = 3$
Example	**What is the value of x and y in this system of equations?** $\begin{cases} 3x - y = 7 \\ -x + 4y = 5 \end{cases}$ **Solution:** Solving System of Equations by Elimination: $\begin{array}{c} 3x - y = 7 \\ \underline{-x + 4y = 5} \end{array}$ Multiply the second equation by 3, then add it to the first equation. $\begin{array}{c} 3x - y = 7 \\ \underline{3(-x + 4y = 5)} \end{array} \Rightarrow \begin{array}{c} 3x - y = 7 \\ \underline{-3x + 12y = 15} \end{array} \Rightarrow 11y = 22 \Rightarrow y = 2.$ Now, substitute 2 for y in the first equation and solve for x. $3x - (2) = 7 \Rightarrow 3x = 9 \Rightarrow x = 3$
Your Turn!	1) $-4x + 4y = 8$ $\quad -4x + 2y = 6$ $\quad x = \_\_\_$ $\quad y = \_\_\_$ 2) $-5x + y = -3$ $\quad 3x - 8y = 24$ $\quad x = \_\_\_$ $\quad y = \_\_\_$ 3) $y = -2$ $\quad 4x - 3y = 8$ $\quad x = \_\_\_$ $\quad y = \_\_\_$ 4) $y = -3x + 5$ $\quad 5x - 4y = -3$ $\quad x = \_\_\_$ $\quad y = \_\_\_$ 5) $20x - 18y = -26$ $\quad -10x + 6y = 22$ $\quad x = \_\_\_$ $\quad y = \_\_\_$ 6) $-9x - 12y = 15$ $\quad 2x - 6y = 14$ $\quad x = \_\_\_$ $\quad y = \_\_\_$

Name:	Date: ....

Topic	System of Equations- Answers
Notes	✓ A system of equations contains two equations and two variables. For example, consider the system of equations: $x - 2y = -2, x + 2y = 10$ ✓ The easiest way to solve a system of equation is using the elimination method. The elimination method uses the addition property of equality. You can add the same value to each side of an equation. ✓ For the first equation above, you can add $x + 2y$ to the left side and 10 to the right side of the first equation: $x - 2y + (x + 2y) = -2 + 10$. Now, if you simplify, you get: $x - 2y + (x + 2y) = -2 + 10 \rightarrow 2x = 8 \rightarrow x = 4$. Now, substitute 4 for the x in the first equation: $4 - 2y = -2$. By solving this equation, $y = 3$
Example	**What is the value of x and y in this system of equations?** $\begin{cases} 3x - y = 7 \\ -x + 4y = 5 \end{cases}$ **Solution:** Solving System of Equations by Elimination: $\begin{array}{c} 3x - y = 7 \\ \hline -x + 4y = 5 \end{array}$ Multiply the second equation by 3, then add it to the first equation. $\begin{array}{c} 3x - y = 7 \\ \hline 3(-x + 4y = 5) \end{array} \Rightarrow \begin{array}{c} 3x - y = 7 \\ \hline -3x + 12y = 15) \end{array} \Rightarrow 11y = 22 \Rightarrow y = 2.$ Now, substitute 2 for y in the first equation and solve for x. $3x - (2) = 7 \Rightarrow 3x = 9 \Rightarrow x = 3$
Your Turn!	1) $-4x + 4y = 8$ $-4x + 2y = 6$ $x = -1$ $y = 1$ 2) $-5x + y = -3$ $3x - 8y = 24$ $x = 0$ $y = -3$ 3) $y = -2$ $4x - 3y = 8$ $x = \dfrac{1}{2}$ $y = -2$ 4) $y = -3x + 5$ $5x - 4y = -3$ $x = 1$ $y = 2$ 5) $20x - 18y = -26$ $-10x + 6y = 22$ $x = -4$ $y = -3$ 6) $-9x - 12y = 15$ $2x - 6y = 14$ $x = 1$ $y = -2$

| Name: .... | Date: .... |

Topic	**Graphing Single–Variable Inequalities**
Notes	✓ An inequality compares two expressions using an inequality sign. ✓ Inequality signs are: "less than" $<$, "greater than" $>$, "less than or equal to" $\leq$, and "greater than or equal to" $\geq$. ✓ To graph a single–variable inequality, find the value of the inequality on the number line. ✓ For less than ($<$) or greater than ($>$) draw open circle on the value of the variable. If there is an equal sign too, then use filled circle. ✓ Draw an arrow to the right for greater or to the left for less than.
Example	***Draw a graph for this inequality.*** $x < 5$ **Solution:** Since, the variable is less than 5, then we need to find 5 in the number line and draw an open circle on it. Then, draw an arrow to the left. -6 -5 -4 -3 -2 -1 0 1 2 3 4 5 6

Your Turn!	1) $x < 4$ -6 -5 -4 -3 -2 -1 0 1 2 3 4 5 6	2) $x \geq -1$ -6 -5 -4 -3 -2 -1 0 1 2 3 4 5 6
	3) $x \geq -3$ -6 -5 -4 -3 -2 -1 0 1 2 3 4 5 6	4) $x \leq 6$ -6 -5 -4 -3 -2 -1 0 1 2 3 4 5 6
	5) $x > -6$ -6 -5 -4 -3 -2 -1 0 1 2 3 4 5 6	6) $2 > x$ -6 -5 -4 -3 -2 -1 0 1 2 3 4 5 6
	7) $-2 \leq x$ -6 -5 -4 -3 -2 -1 0 1 2 3 4 5 6	8) $x > 0$ -6 -5 -4 -3 -2 -1 0 1 2 3 4 5 6

Name:	Date: ....

Topic	Graphing Single–Variable Inequalities- Answers
Notes	✓ An inequality compares two expressions using an inequality sign. ✓ Inequality signs are: "less than" <, "greater than" >, "less than or equal to" ≤, and "greater than or equal to" ≥. ✓ To graph a single–variable inequality, find the value of the inequality on the number line. ✓ For less than (<) or greater than (>) draw open circle on the value of the variable. If there is an equal sign too, then use filled circle. ✓ Draw an arrow to the right for greater or to the left for less than.
Example	**Draw a graph for this inequality.** $x < 5$ **Solution:** Since, the variable is less than 5, then we need to find 5 in the number line and draw an open circle on it. Then, draw an arrow to the left. ‹————————————○——› -6 -5 -4 -3 -2 -1 0 1 2 3 4 5 6
Your Turn!	1) $x < 4$ ‹————————○——› -6 -5 -4 -3 -2 -1 0 1 2 3 4 5 6 3) $x \geq -3$ ‹——●————————› -6 -5 -4 -3 -2 -1 0 1 2 3 4 5 6 5) $x > -6$ ‹○————————————› -6 -5 -4 -3 -2 -1 0 1 2 3 4 5 6 7) $-2 \leq x$ ‹————●————————› -6 -5 -4 -3 -2 -1 0 1 2 3 4 5 6	2) $x \geq -1$ ‹————●——————› -6 -5 -4 -3 -2 -1 0 1 2 3 4 5 6 4) $x \leq 6$ ‹————————————●› -6 -5 -4 -3 -2 -1 0 1 2 3 4 5 6 6) $2 > x$ ‹——————————○——› -6 -5 -4 -3 -2 -1 0 1 2 3 4 5 6 8) $x > 0$ ‹——————○————————› -6 -5 -4 -3 -2 -1 0 1 2 3 4 5 6

Name:	Date: ....

Topic	**One–Step Inequalities**
Notes	✓ Inequality signs are: "less than" $<$, "greater than" $>$, "less than or equal to" $\leq$, and "greater than or equal to" $\geq$. ✓ You only need to perform one Math operation in order to solve the one-step inequalities. ✓ To solve one-step inequalities, find the inverse (opposite) operation is being performed. ✓ For dividing or multiplying both sides by negative numbers, flip the direction of the inequality sign.
Example	**Solve this inequality.** $x + 12 < 60 \Rightarrow$ _____ Here, the operation is addition and its inverse operation is subtraction. To solve this inequality, subtract 12 from both sides of the ***inequality:*** $x + 12 - 12 < 60 - 12$ Then simplify: $x < 48$
Your Turn!	1) $4x < -8 \Rightarrow$ _____ 2) $x + 6 > 28 \Rightarrow$ _____ 3) $-3x \geq 36 \Rightarrow$ _____ 4) $x - 16 \leq 4 \Rightarrow$ _____ 5) $\frac{x}{2} \geq -9 \Rightarrow$ _____ 6) $48 < 6x \Rightarrow$ _____ 7) $77 \leq 11x \Rightarrow$ _____ 8) $\frac{x}{4} > 9 \Rightarrow$ _____

Name: ....

Date: ....

Topic	One–Step Inequalities - Answers
Notes	✓ Inequality signs are: "less than" $<$, "greater than" $>$, "less than or equal to" $\leq$, and "greater than or equal to" $\geq$. ✓ You only need to perform one Math operation in order to solve the one-step inequalities. ✓ To solve one-step inequalities, find the inverse (opposite) operation is being performed. ✓ For dividing or multiplying both sides by negative numbers, flip the direction of the inequality sign.
Example	**Solve this inequality.** $x + 12 < 60 \Rightarrow$ _____ Here, the operation is addition and its inverse operation is subtraction. To solve this inequality, subtract 12 from both sides of the inequality: $x + 12 - 12 < 60 - 12$ Then simplify: $x < 48$

Your Turn!

1) $4x < -8 \Rightarrow x < -2$	2) $x + 6 > 28 \Rightarrow x > 22$
3) $-3x \geq 36 \Rightarrow x \leq -12$	4) $x - 16 \leq 4 \Rightarrow x \leq 20$
5) $\frac{x}{2} \geq -9 \Rightarrow x \geq -18$	6) $48 < 6x \Rightarrow 8 < x$
7) $77 \leq 11x \Rightarrow 7 \leq x$	8) $\frac{x}{4} > 9 \Rightarrow x > 36$

Name:	Date: ....

Topic	**Multi –Step Inequalities**	
Notes	✓ Isolate the variable. ✓ Simplify using the inverse of addition or subtraction. ✓ Simplify further by using the inverse of multiplication or division. ✓ For dividing or multiplying both sides by negative numbers, flip the direction of the inequality sign.	
Example	*Solve this inequality.* $3x + 12 \leq 21$ **Solution:** First subtract 12 from both sides: $3x + 12 - 12 \leq 21 - 12$ Then simplify: $3x + 12 - 12 \leq 21 - 12 \rightarrow 3x \leq 9$ Now divide both sides by 3: $\frac{3x}{3} \leq \frac{9}{3} \rightarrow x \leq 3$	
Your Turn!	1) $5x + 6 < 36 \rightarrow$ _____	2) $2x - 8 \leq 6 \rightarrow$ _____
	3) $2x - 5 \leq 17 \rightarrow$ _____	4) $14 - 7x \geq -7 \rightarrow$ _____
	5) $18 - 6x \geq -6 \rightarrow$ _____	6) $2x - 18 \leq 16 \rightarrow$ _____
	7) $8 + 4x < 44 \rightarrow$ _____	8) $5 - 4x < 17 \rightarrow$ _____

Name:	Date: ....

Topic	**Multi –Step Inequalities - Answers**
Notes	✓ Isolate the variable. ✓ Simplify using the inverse of addition or subtraction. ✓ Simplify further by using the inverse of multiplication or division. ✓ For dividing or multiplying both sides by negative numbers, flip the direction of the inequality sign.

Example	**Solve this inequality. $3x + 12 \leq 21$**
	   Solution: First subtract 12 from both sides: $3x + 12 - 12 \leq 21 - 12$
	   Then simplify: $3x + 12 - 12 \leq 21 - 12 \rightarrow 3x \leq 9$
	   Now divide both sides by 3: $\frac{3x}{3} \leq \frac{9}{3} \rightarrow x \leq 3$

Your Turn!	1) $5x + 6 < 36 \rightarrow x < 6$ 2) $2x - 8 \leq 6 \rightarrow x \leq 7$ 3) $2x - 5 \leq 17 \rightarrow x \leq 11$ 4) $14 - 7x \geq -7 \rightarrow x \leq 3$ 5) $18 - 6x \geq -6 \rightarrow x \leq 4$ 6) $2x - 18 \leq 16 \rightarrow x \leq 17$ 7) $8 + 4x < 44 \rightarrow x < 9$ 8) $5 - 4x < 17 \rightarrow x > -3$

Name:	Date: ....

Topic	**Finding Slope**
Notes	✓ The slope of a line represents the direction of a line on the coordinate plane. ✓ A line on coordinate plane can be drawn by connecting two points. ✓ To find the slope of a line, we need two points. ✓ The slope of a line with two points A (x_1, y_1) and B (x_2, y_2) can be found by using this formula: $\frac{y_2 - y_1}{x_2 - x_1} = \frac{rise}{run}$ ✓ The equation of a line is typically written as $y = mx + b$ where m is the slope and b is the y-intercept.
Examples	**Find the slope of the line through these two points: $(4, -12)$ and $(9, 8)$.** **Solution:** Slope $= \frac{y_2 - y_1}{x_2 - x_1}$. Let (x_1, y_1) be $(4, -12)$ and (x_2, y_2) be $(9, 8)$. Then: slope $= \frac{y_2 - y_1}{x_2 - x_1} = \frac{8-(-12)}{9-4} = \frac{8+12}{5} = \frac{20}{5} = 4$ **Find the slope of the line with equation $y = 5x - 6$** **Solution:** when the equation of a line is written in the form of $y = mx + b$, the slope is m. In this line: $y = 5x - 6$, the slope is 5.

Your Turn!	1) $(2, 3), (4, 7)$
	
	   Slope = ____
	
	 2) $(-2, 2), (0, 4)$
	
	   Slope = ____
	
	 3) $(4, -2), (2, 4)$
	
	   Slope = ____
	
	 4) $(-4, -1), (0, 7)$
	
	   Slope = ____
	
	 5) $y = 3x + 18$
	
	   Slope = ____
	
	 6) $y = 12x - 3$
	
	   Slope = ____

Name:	Date: ....

Topic	**Finding Slope - Answers**
Notes	✓ The slope of a line represents the direction of a line on the coordinate plane. ✓ A line on coordinate plane can be drawn by connecting two points. ✓ To find the slope of a line, we need two points. ✓ The slope of a line with two points A (x_1, y_1) and B (x_2, y_2) can be found by using this formula: $\frac{y_2 - y_1}{x_2 - x_1} = \frac{rise}{run}$ ✓ The equation of a line is typically written as $y = mx + b$ where m is the slope and b is the y-intercept.
Examples	**Find the slope of the line through these two points: $(4, -12)$ and $(9, 8)$.** **Solution:** Slope $= \frac{y_2 - y_1}{x_2 - x_1}$. Let (x_1, y_1) be $(4, -12)$ and (x_2, y_2) be $(9, 8)$. Then: slope $= \frac{y_2 - y_1}{x_2 - x_1} = \frac{8 - (-12)}{9 - 4} = \frac{8 + 12}{5} = \frac{20}{5} = 4$ **Find the slope of the line with equation $y = 5x - 6$** **Solution:** when the equation of a line is written in the form of $y = mx + b$, the slope is m. In this line: $y = 5x - 6$, the slope is 5.
Your Turn!	1) $(2, 3), (4, 7)$ Slope $= 2$	2) $(-2, 2), (0, 4)$ Slope $= 1$
	3) $(4, -2), (2, 4)$ Slope $= -3$	4) $(-4, -1), (0, 7)$ Slope $= 2$
	5) $y = 3x + 18$ Slope $= 3$	6) $y = 12x - 3$ Slope $= 12$

Name: ....   **Date:** ....

Topic	Graphing Lines Using Slope–Intercept Form
Notes	✓ Slope–intercept form of a line: given the slope m and the y–intercept (the intersection of the line and y-axis) b, then the equation of the line is: $$y = mx + b$$
Example	**Sketch the graph of** $y = -2x - 1$. **Solution:** To graph this line, we need to find two points. When x is zero the value of y is -1. And when y is zero the value of x is $-\frac{1}{2}$. $x = 0 \rightarrow y = -2(0) - 1 \rightarrow y = -1$, $y = 0 \rightarrow 0 = -2x - 1 \rightarrow x = -\frac{1}{2}$ Now, we have two points: $(0, -1)$ and $(-\frac{1}{2}, 0)$. Find the points and graph the line. Remember that the slope of the line is $-\frac{1}{2}$.
Your Turn!	1) $y = -4x + 1$ 2) $y = -x - 5$

Name:	Date: ....

Topic	**Graphing Lines Using Slope–Intercept Form - Answers**
Notes	✓ Slope–intercept form of a line: given the slope m and the y–intercept (the intersection of the line and y-axis) b, then the equation of the line is: $$y = mx + b$$
Example	**Sketch the graph of** $y = -2x - 1$. **Solution:** To graph this line, we need to find two points. When x is zero the value of y is -1. And when y is zero the value of x is $-\frac{1}{2}$. $$x = 0 \rightarrow y = -2(0) - 1 \rightarrow y = -1$$ $$y = 0 \rightarrow 0 = -2x - 1 \rightarrow x = -\frac{1}{2}$$ Now, we have two points: $(0, -1)$ and $(-\frac{1}{2}, 0)$. Find the points and graph the line. Remember that the slope of the line is $-\frac{1}{2}$.
Your Turn!	1) $y = -4x + 1$ 2) $y = -x - 5$

Name:	Date: ....

Topic	Writing Linear Equations
Notes	✓ The equation of a line: $y = mx + b$ ✓ Identify the slope. ✓ Find the y–intercept. This can be done by substituting the slope and the coordinates of a point (x, y) on the line.
Example	**Write the equation of the line through** $(3, 1)$ **and** $(-1, 5)$. **Solution:** $Slop = \frac{y_2 - y_1}{x_2 - x_1} = \frac{5-1}{-1-3} = \frac{4}{-4} = -1 \rightarrow m = -1$ To find the value of b, you can use either points. The answer will be the same: $y = -x + b$ $(3, 1) \rightarrow 1 = -3 + b \rightarrow b = 4$ $(-1, 5) \rightarrow 5 = -(-1) + b \rightarrow b = 4$ The equation of the line is: $y = -x + 4$
Your Turn!	1) through: $(-2, 7), (1, 4)$ $y =$ 2) through: $(6, 1), (5, 2)$ $y =$ 3) through: $(5, -1), (8, 2)$ $y =$ 4) through: $(-2, 4), (4, -8)$ $y =$ 5) through: $(6, -5), (-5, 6)$ $y =$ 6) through: $(4, -4), (-2, 8)$ $y =$ 7) through $(8, 8)$, Slope: 2 $y =$ 8) through $(-7, 10)$, Slope: -2 $y =$

Name:	Date: ....

Topic	**Writing Linear Equations - Answers**
Notes	✓ The equation of a line: $y = mx + b$ ✓ Identify the slope. ✓ Find the y–intercept. This can be done by substituting the slope and the coordinates of a point (x, y) on the line.
Example	**Write the equation of the line through $(3, 1)$ and $(-1, 5)$.** **Solution:** $Slop = \frac{y_2 - y_1}{x_2 - x_1} = \frac{5-1}{-1-3} = \frac{4}{-4} = -1 \rightarrow m = -1$ To find the value of b, you can use either points. The answer will be the same: $y = -x + b$ $(3, 1) \rightarrow 1 = -3 + b \rightarrow b = 4$ $(-1, 5) \rightarrow 5 = -(-1) + b \rightarrow b = 4$ The equation of the line is: $y = -x + 4$

Your Turn!	1) through: $(-2, 7), (1, 4)$ $y = -x + 5$	2) through: $(6, 1), (5, 2)$ $y = -x + 7$
	3) through: $(5, -1), (8, 2)$ $y = x - 6$	4) through: $(-2, 4), (4, -8)$ $y = -2x$
	5) through: $(6, -5), (-5, 6)$ $y = -x + 1$	6) through: $(4, -4), (-2, 8)$ $y = -2x + 4$
	7) through $(8, 8)$, Slope: 2 $y = 2x - 8$	8) through $(-7, 10)$, Slope: -2 $y = -2x - 4$

Name:	Date: ....

Topic	**Finding Midpoint**
Notes	✓ The middle of a line segment is its midpoint. ✓ The Midpoint of two endpoints A (x_1, y_1) and B (x_2, y_2) can be found using this formula: $M\left(\frac{x_1+x_2}{2}, \frac{y_1+y_2}{2}\right)$
Example	**Find the midpoint of the line segment with the given endpoints.** $(1, -2), (3, 6)$ **Solution:** Midpoint $= \left(\frac{x_1+x_2}{2}, \frac{y_1+y_2}{2}\right) \rightarrow (x_1, y_1) = (1, -2)$ and $(x_2, y_2) = (3, 6)$ Midpoint $= \left(\frac{1+3}{2}, \frac{-2+6}{2}\right) \rightarrow \left(\frac{4}{2}, \frac{4}{2}\right) \rightarrow M(2, 2)$

Your Turn!	1) $(6, 0), (-4, 2)$ Midpoint = (___, ___)	2) $(4, -1), (2, 3)$ Midpoint = (___, ___)
	3) $(-3, 4), (-5, 0)$ Midpoint = (___, ___)	4) $(8, 1), (-4, 5)$ Midpoint = (___, ___)
	5) $(6, 7), (-4, 5)$ Midpoint = (___, ___)	6) $(2, -3), (2, 5)$ Midpoint = (___, ___)
	7) $(7, 3), (-1, -7)$ Midpoint = (___, ___)	8) $(3, 9), (-1, 5)$ Midpoint = (___, ___)
	9) $(3, 4), (-7, -6)$ Midpoint = (___, ___)	10) $(-5, 2), (11, -6)$ Midpoint = (___, ___)

| Name: .... | Date: .... |

Topic	**Finding Midpoint - Answers**
Notes	✓ The middle of a line segment is its midpoint. ✓ The Midpoint of two endpoints A (x_1, y_1) and B (x_2, y_2) can be found using this formula: $M\left(\frac{x_1+x_2}{2}, \frac{y_1+y_2}{2}\right)$
Example	**Find the midpoint of the line segment with the given endpoints.** $(1, -2), (3, 6)$ **Solution:** Midpoint $= \left(\frac{x_1+x_2}{2}, \frac{y_1+y_2}{2}\right) \to (x_1, y_1) = (1, -2)$ and $(x_2, y_2) = (3, 6)$ Midpoint $= \left(\frac{1+3}{2}, \frac{-2+6}{2}\right) \to \left(\frac{4}{2}, \frac{4}{2}\right) \to M(2, 2)$

Your Turn!	1) $(6, 0), (-4, 2)$ Midpoint $= (1, 1)$	2) $(4, -1), (2, 3)$ Midpoint $= (3, 1)$
	3) $(-3, 4), (-5, 0)$ Midpoint $= (-4, 2)$	4) $(8, 1), (-4, 5)$ Midpoint $= (2, 3)$
	5) $(6, 7), (-4, 5)$ Midpoint $= (1, 6)$	6) $(2, -3), (2, 5)$ Midpoint $= (2, 1)$
	7) $(7, 3), (-1, -7)$ Midpoint $= (3, -2)$	8) $(3, 9), (-1, 5)$ Midpoint $= (1, 7)$
	9) $(3, 4), (-7, -6)$ Midpoint $= (-2, -1)$	10) $(-5, 2), (11, -6)$ Midpoint $= (3, -2)$

Name:	Date: ....

Topic	**Finding Distance of Two Points**
Notes	✓ Use this formula to find the distance of two points A (x_1, y_1) and B (x_2, y_2): $$d = \sqrt{(x_2 - x_1)^2 + (y_2 - y_1)^2}$$
Example	**Find the distance of two points $(-1, 5)$ and $(4, -7)$.** **Solution:** Use distance of two points formula: $d = \sqrt{(x_2 - x_1)^2 + (y_2 - y_1)^2}$ $(x_1, y_1) = (-1, 5)$, and $(x_2, y_2) = (4, -7)$ Then: $d = \sqrt{(x_2 - x_1)^2 + (y_2 - y_1)^2} \rightarrow$ $d = \sqrt{(4 - (-1))^2 + (-7 - 5)^2} =$ $\sqrt{(5)^2 + (-12)^2} = \sqrt{25 + 144} = \sqrt{169} = 13$
Your Turn!	1) $(6, 2), (-4, 2)$ Distance = ____ 3) $(-5, 10), (7, 1)$ Distance = ____ 5) $(-3, 6), (-4, 5)$ Distance = ____ 7) $(-3, 4), (-5, 0)$ Distance = ____ 2) $(2, -3), (2, 5)$ Distance = ____ 4) $(8, 1), (-4, 6)$ Distance = ____ 6) $(4, -1), (14, 23)$ Distance = ____ 8) $(3, 9), (-1, 5)$ Distance = ____

Name:	**Date:** ....

Topic	Finding Distance of Two Points - Answers
Notes	✓ Use this formula to find the distance of two points A (x_1, y_1) and B (x_2, y_2): $$d = \sqrt{(x_2 - x_1)^2 + (y_2 - y_1)^2}$$
Example	**Find the distance of two points** $(-1, 5)$ **and** $(4, -7)$. **Solution:** Use distance of two points formula: $d = \sqrt{(x_2 - x_1)^2 + (y_2 - y_1)^2}$ $(x_1, y_1) = (-1, 5)$, and $(x_2, y_2) = (4, -7)$ Then: $d = \sqrt{(x_2 - x_1)^2 + (y_2 - y_1)^2} \rightarrow$ $d = \sqrt{(4 - (-1))^2 + (-7 - 5)^2} =$ $\sqrt{(5)^2 + (-12)^2} = \sqrt{25 + 144} = \sqrt{169} = 13$

Your Turn!	1) $(6, 2), (-4, 2)$ Distance $= 10$	2) $(2, -3), (2, 5)$ Distance $= 8$
	3) $(-5, 10), (7, 1)$ Distance $= 15$	4) $(8, 1), (-4, 6)$ Distance $= 13$
	5) $(-3, 6), (-4, 5)$ Distance $= \sqrt{2}$	6) $(4, -1), (14, 23)$ Distance $= 26$
	7) $(-3, 4), (-5, 0)$ Distance $= \sqrt{20} = 2\sqrt{5}$	8) $(3, 9), (-1, 5)$ Distance $= \sqrt{32} = 4\sqrt{2}$

Name: .... Date: ....

Topic	Graphing Linear Inequalities
Notes	✓ To graph a linear inequality, first draw a graph of the "equals" line. ✓ Choose a testing point. (it can be any point on both sides of the line.) ✓ Put the value of (x, y) of that point in the inequality. If that works, that part of the line is the solution. If the values don't work, then the other part of the line is the solution.
Example	**Sketch the graph of** $y < 3x + 2$ **Solution:** First, graph the line: $y = 3x + 2$ The slope is 3 and y-intercept is 2. Then, choose a testing point. The easiest point to test is the origin: $(0, 0)$ $(0,0) \rightarrow y < 3x + 2 \rightarrow 0 < 3(0) + 2 \rightarrow 0 < 2$ 0 is less than 2. So, this part of the line (on the right side) is the solution.
Your Turn!	1) $y > 4x + 2$ 2) $y < -2x + 5$

Name:	Date: ....

Topic	**Graphing Linear Inequalities - Answers**
Notes	✓ To graph a linear inequality, first draw a graph of the "equals" line. ✓ Choose a testing point. (it can be any point on both sides of the line.) ✓ Put the value of (x, y) of that point in the inequality. If that works, that part of the line is the solution. If the values don't work, then the other part of the line is the solution.
Example	**Sketch the graph of** $y < 3x + 2$ **Solution:** First, graph the line: $y = 3x + 2$ The slope is 3 and y-intercept is 2. Then, choose a testing point. The easiest point to test is the origin: $(0, 0)$ $(0,0) \to y < 3x + 2 \to 0 < 3(0) + 2 \to 0 < 2$ 0 is less than 2. So, this part of the line (on the right side) is the solution.
Your Turn!	1) $y > 4x + 2$ 2) $y < -2x + 5$

Name:	Date: ....

Topic	**Multiplication Property of Exponents**
Notes	✓ Exponents are shorthand for repeated multiplication of the same number by itself. For example, instead of 2×2, we can write 2^2. For $3 \times 3 \times 3 \times 3$, we can write 3^4 ✓ In algebra, a variable is a letter used to stand for a number. The most common letters are: $x, y, z, a, b, c, m,$ and n. ✓ Exponent's rules: $x^a \times x^b = x^{a+b}$, $\frac{x^a}{x^b} = x^{a-b}$ $(x^a)^b = x^{a \times b}$ $\qquad$ $(xy)^a = x^a \times y^a$ $\qquad$ $\left(\frac{a}{b}\right)^c = \frac{a^c}{b^c}$
Example	**Multiply.** $4x^3 \times 2x^2$ Use Exponent's rules: $x^a \times x^b = x^{a+b} \rightarrow x^3 \times x^2 = x^{3+2} = x^5$ Then: $4x^3 \times 2x^2 = 8x^5$

Your Turn!	1) $x^2 \times 3x =$	2) $5x^4 \times x^2 =$
	3) $3x^2 \times 4x^5 =$	4) $3x^2 \times 6xy =$
	5) $3x^5y \times 5x^2y^3 =$	6) $3x^2y^2 \times 5x^2y^8 =$
	7) $5x^2y \times 5x^2y^7 =$	8) $6x^6 \times 4x^9y^4 =$
	9) $8x^2y^5 \times 7x^5y^3 =$	10) $12x^6x^2 \times 3xy^5 =$

Name:	Date: ....

Topic	Multiplication Property of Exponents - Answers
Notes	✓ Exponents are shorthand for repeated multiplication of the same number by itself. For example, instead of 2×2, we can write 2^2. For $3 \times 3 \times 3 \times 3$, we can write 3^4 ✓ In algebra, a variable is a letter used to stand for a number. The most common letters are: $x, y, z, a, b, c, m,$ and n. ✓ Exponent's rules: $x^a \times x^b = x^{a+b}$, $\dfrac{x^a}{x^b} = x^{a-b}$ $\quad (x^a)^b = x^{a \times b} \qquad\qquad (xy)^a = x^a \times y^a \qquad \left(\dfrac{a}{b}\right)^c = \dfrac{a^c}{b^c}$
Example	**Multiply.** $4x^3 \times 2x^2$ Use Exponent's rules: $x^a \times x^b = x^{a+b} \rightarrow x^3 \times x^2 = x^{3+2} = x^5$ Then: $4x^3 \times 2x^2 = 8x^5$

Your Turn!	1) $x^2 \times 3x = 3x^3$	2) $5x^4 \times x^2 = 5x^6$
	3) $3x^2 \times 4x^5 = 12x^7$	4) $3x^2 \times 6xy = 18x^3y$
	5) $3x^5y \times 5x^2y^3 = 15x^7y^4$	6) $3x^2y^2 \times 5x^2y^8 = 15x^4y^{10}$
	7) $5x^2y \times 5x^2y^7 = 25x^4y^8$	8) $6x^6 \times 4x^9y^4 = 24x^{15}y^4$
	9) $8x^2y^5 \times 7x^5y^3 = 56x^7y^8$	10) $12x^6x^2 \times 3xy^5 = 36x^9y^5$

| Name: .... | Date: .... |

Topic	**Division Property of Exponents**
Notes	✓ For division of exponents use these formulas: $\frac{x^a}{x^b} = x^{a-b}$, $x \neq 0$ $\frac{x^a}{x^b} = \frac{1}{x^{b-a}}$, $x \neq 0$, $\qquad$ $\frac{1}{x^b} = x^{-b}$
Example	**Simplify.** $\frac{6x^3y}{36x^2y^3}$ First cancel the common factor: $6 \rightarrow \frac{6x^3y}{36x^2y^3} = \frac{x^3y}{6x^2y^3}$ Use Exponent's rules: $\frac{x^a}{x^b} = x^{a-b} \rightarrow \frac{x^3}{x^2} = x^{3-2} = x^1 = x$ Then: $\frac{6x^3y}{36x^2y^3} = \frac{xy}{6y^3} \rightarrow$ now cancel the common factor: $y \rightarrow \frac{xy}{6y^3} = \frac{x}{6y^2}$

Your Turn!	1) $\frac{3^7}{3^2} =$	2) $\frac{5x}{10x^3} =$
	3) $\frac{3x^3}{2x^5} =$	4) $\frac{12x^3}{14x^6} =$
	5) $\frac{12x^3}{9y^8} =$	6) $\frac{25xy^4}{5x^6y^2} =$
	7) $\frac{2x^4y^5}{7xy^2} =$	8) $\frac{16x^2y^8}{4x^3} =$
	9) $\frac{12x^4}{15x^7y^9} =$	10) $\frac{12yx^4}{10yx^8} =$

| Name: .... | Date: .... |

Topic	**Division Property of Exponents - Answers**
Notes	✓ For division of exponents use following formulas: $\frac{x^a}{x^b} = x^{a-b}$, $x \neq 0$ $\frac{x^a}{x^b} = \frac{1}{x^{b-a}}$, $x \neq 0$,　　　$\frac{1}{x^b} = x^{-b}$
Example	**Simplify.** $\frac{6x^3 y}{36x^2 y^3}$ First cancel the common factor: $6 \rightarrow \frac{6x^3 y}{36x^2 y^3} = \frac{x^3 y}{6x^2 y^3}$ Use Exponent's rules: $\frac{x^a}{x^b} = x^{a-b} \rightarrow \frac{x^3}{x^2} = x^{3-2} = x^1 = x$ Then: $\frac{6x^3 y}{36x^2 y^3} = \frac{xy}{6y^3} \rightarrow$ now cancel the common factor: $y \rightarrow \frac{xy}{6y^3} = \frac{x}{6y^2}$

Your Turn!	1) $\frac{3^7}{3^2} = 3^5$	2) $\frac{5x}{10x^3} = \frac{1}{2x^2}$
	3) $\frac{3x^3}{2x^5} = \frac{3}{2x^2}$	4) $\frac{12x^3}{14x^6} = \frac{6}{7x^3}$
	5) $\frac{12x^3}{9y^8} = \frac{4x^3}{3y^8}$	6) $\frac{25xy^4}{5x^6 y^2} = \frac{5y^2}{x^5}$
	7) $\frac{2x^4 y^5}{7xy^2} = \frac{2x^3 y^3}{7}$	8) $\frac{16x^2 y^8}{4x^3} = \frac{4y^8}{x}$
	9) $\frac{12x^4}{15x^7 y^9} = \frac{4}{5x^3 y^9}$	10) $\frac{12y^8 x^4}{10y^2 x^8} = \frac{6}{5x^4}$

Name:	Date: ....

Topic	**Powers of Products and Quotients**	
Notes	✓ For any nonzero numbers a and b and any integer x, $$(ab)^x = a^x \times b^x, \left(\frac{a}{b}\right)^c = \frac{a^c}{b^c}$$	
Example	**Simplify.** $\left(\frac{2x^3}{x}\right)^2$ First cancel the common factor: $x \rightarrow \left(\frac{2x^3}{x}\right)^2 = (2x^2)^2$ Use Exponent's rules: $(ab)^x = a^x \times b^x$ Then: $(2x^2)^2 = (2)^2(x^2)^2 = 4x^4$	
Your Turn!	1) $(4x^3x^3)^2 =$	2) $(3x^3 \times 5x)^2 =$
	3) $(10x^{11}y^3)^2 =$	4) $(9x^7y^5)^2 =$
	5) $(4x^4y^6)^3 =$	6) $(3x \times 4y^3)^2 =$
	7) $\left(\frac{5x}{x^2}\right)^2 =$	8) $\left(\frac{x^4y^4}{x^2y^2}\right)^3 =$
	9) $\left(\frac{25x}{5x^6}\right)^2 =$	10) $\left(\frac{x^8}{x^6y^2}\right)^2 =$

Name:	Date: ....

Topic	**Powers of Products and Quotients - Answers**
Notes	✓ For any nonzero numbers a and b and any integer x, $$(ab)^x = a^x \times b^x, \left(\frac{a}{b}\right)^c = \frac{a^c}{b^c}$$
Example	**Simplify.** $\left(\frac{2x^3}{x}\right)^2$ First cancel the common factor: $x \rightarrow \left(\frac{2x^3}{x}\right)^2 = (2x^2)^2$ Use Exponent's rules: $(ab)^x = a^x \times b^x$ Then: $(2x^2)^2 = (2)^2(x^2)^2 = 4x^4$
Your Turn!	1) $(4x^3x^3)^2 = 16x^{12}$ 2) $(3x^3 \times 5x)^2 = 225x^8$ 3) $(10x^{11}y^3)^2 = 100x^{22}y^6$ 4) $(9x^7y^5)^2 = 81x^{14}y^{10}$ 5) $(4x^4y^6)^3 = 64x^{12}y^{18}$ 6) $(3x \times 4y^3)^2 = 144x^2y^6$ 7) $\left(\frac{5x}{x^2}\right)^2 = \frac{25}{x^2}$ 8) $\left(\frac{x^4y^4}{x^2y^2}\right)^3 = x^6y^6$ 9) $\left(\frac{25x}{5x^6}\right)^2 = \frac{25}{x^{10}}$ 10) $\left(\frac{x^8}{x^6y^2}\right)^2 = \frac{x^4}{y^4}$

| Name: .... | Date: .... |

Topic	**Zero and Negative Exponents**
Notes	✓ A negative exponent is the reciprocal of that number with a positive exponent. $(3)^{-2} = \frac{1}{3^2}$ ✓ Zero-Exponent Rule: $a^0 = 1$, this means that anything raised to the zero power is 1. For example: $(28x^2y)^0 = 1$
Example	**Evaluate.** $\left(\frac{1}{3}\right)^{-2} =$ Use negative exponent's rule: $\left(\frac{1}{x^a}\right)^{-2} = (x^a)^2 \rightarrow \left(\frac{1}{3}\right)^{-2} = (3)^2$ Then: $(3)^2 = 9$

Your Turn!	1) $2^{-3} =$	2) $3^{-3} =$
	3) $7^{-3} =$	4) $1^{-3} =$
	5) $8^{-3} =$	6) $4^{-4} =$
	7) $10^{-3} =$	8) $7^{-4} =$
	9) $\left(\frac{1}{8}\right)^{-1} =$	10) $\left(\frac{1}{5}\right)^{-2} =$

Name:	Date: ....

Topic	Zero and Negative Exponents - Answers
Notes	✓ A negative exponent is the reciprocal of that number with a positive exponent. $(3)^{-2} = \frac{1}{3^2}$ ✓ Zero-Exponent Rule: $a^0 = 1$, this means that anything raised to the zero power is 1. For example: $(28x^2y)^0 = 1$
Example	**Evaluate.** $\left(\frac{1}{3}\right)^{-2} =$ Use negative exponent's rule: $\left(\frac{1}{x^a}\right)^{-2} = (x^a)^2 \rightarrow \left(\frac{1}{3}\right)^{-2} = (3)^2$ Then: $(3)^2 = 9$

Your Turn!	1) $2^{-3} = \frac{1}{8}$	2) $3^{-3} = \frac{1}{27}$
	3) $7^{-3} = \frac{1}{343}$	4) $1^{-3} = 1$
	5) $8^{-3} = \frac{1}{512}$	6) $4^{-4} = \frac{1}{256}$
	7) $10^{-3} = \frac{1}{1,000}$	8) $7^{-4} = \frac{1}{2,401}$
	9) $\left(\frac{1}{8}\right)^{-1} = 8$	10) $\left(\frac{1}{5}\right)^{-2} = 25$

Name:	Date: ....

Topic	**Negative Exponents and Negative Bases**	
Notes	✓ Make the power positive. A negative exponent is the reciprocal of that number with a positive exponent. ✓ The parenthesis is important! 5^{-2} is not the same as $(-5)^{-2}$ $$(-5)^{-2} = -\frac{1}{5^2} \text{ and } (-5)^{-2} = +\frac{1}{5^2}$$	
Example	**Simplify.** $\left(-\frac{3x}{4yz}\right)^{-3} =$ Use negative exponent's rule: $\left(\frac{x^a}{x^b}\right)^{-2} = \left(\frac{x^b}{x^a}\right)^2 \rightarrow \left(-\frac{3x}{4yz}\right)^{-3} = \left(-\frac{4yz}{3x}\right)^3$ Now use exponent's rule: $\left(\frac{a}{b}\right)^c = \frac{a^c}{b^c} \rightarrow \left(-\frac{4yz}{3x}\right)^3 = \frac{4^3 y^3 z^3}{3^3 x^3} = \frac{64 y^3 z^3}{27 x^3}$	
Your Turn!	1) $-5x^{-2}y^{-3} =$	2) $20x^{-4}y^{-1} =$
	3) $14a^{-6}b^{-7} =$	4) $-12x^2 y^{-3} =$
	5) $-\frac{25}{x^{-6}} =$	6) $\frac{7b}{-9c^{-4}} =$
	7) $\frac{7ab}{a^{-3}b^{-1}} =$	8) $-\frac{5n^{-2}}{10p^{-3}} = -$
	9) $\frac{4ab^{-2}}{-3c^{-2}} =$	10) $\left(\frac{3a}{2c}\right)^{-2} =$

Name: ....   **Date:** ....

Topic	Negative Exponents and Negative Bases - Answers

Notes

✓ Make the power positive. A negative exponent is the reciprocal of that number with a positive exponent.
✓ The parenthesis is important!
5^{-2} is not the same as $(-5)^{-2}$

$$(-5)^{-2} = -\frac{1}{5^2} \text{ and } (-5)^{-2} = +\frac{1}{5^2}$$

Example

Simplify. $\left(-\frac{3x}{4yz}\right)^{-3} =$

Use negative exponent's rule: $\left(\frac{x^a}{x^b}\right)^{-2} = \left(\frac{x^b}{x^a}\right)^2 \rightarrow \left(-\frac{3x}{4yz}\right)^{-3} = \left(-\frac{4yz}{3x}\right)^3$

Now use exponent's rule: $\left(\frac{a}{b}\right)^c = \frac{a^c}{b^c} \rightarrow \left(-\frac{4yz}{3x}\right)^3 = \frac{4^3 y^3 z^3}{3^3 x^3} = \frac{64 y^3 z^3}{27 x^3}$

Your Turn!

1) $-5x^{-2}y^{-3} = -\dfrac{5}{x^2 y^3}$

2) $20x^{-4}y^{-1} = \dfrac{20}{x^4 y}$

3) $14a^{-6}b^{-7} = \dfrac{14}{a^6 b^7}$

4) $-12x^2 y^{-3} = -\dfrac{12x^2}{y^3}$

5) $-\dfrac{25}{x^{-6}} = -25x^6$

6) $\dfrac{7b}{-9c^{-4}} = -\dfrac{7bc^4}{9}$

7) $\dfrac{7ab}{a^{-3}b^{-1}} = 7a^4 b^2$

8) $-\dfrac{5n^{-2}}{10p^{-3}} = -\dfrac{p^3}{2n^2}$

9) $\dfrac{4ab^{-2}}{-3c^{-2}} = -\dfrac{4ac^2}{3b^2}$

10) $\left(\dfrac{3a}{2c}\right)^{-2} = \dfrac{4c^2}{9a^2}$

Name:	Date: ....

Topic	**Scientific Notation**	
Notes	✓ It is used to write very big or very small numbers in decimal form. ✓ In scientific notation all numbers are written in the form of: $$m \times 10^n$$ <table><tr><td>**Decimal notation**</td><td>**Scientific notation**</td></tr><tr><td>3</td><td>3×10^0</td></tr><tr><td>$-45,000$</td><td>-4.5×10^4</td></tr><tr><td>0.3</td><td>3×10^{-1}</td></tr><tr><td>$2,122.456$</td><td>2.122456×10^3</td></tr></table>	
Example	**Write 0.00054 in scientific notation.** First, move the decimal point to the right so that you have a number that is between 1 and 10. Then: $m = 5.4$ Now, determine how many places the decimal moved in step 1 by the power of 10. Then: $10^{-4} \rightarrow$ When the decimal moved to the right, the exponent is negative. Then: $0.00054 = 5.4 \times 10^{-4}$	
Your Turn!	1) $0.000325 =$	2) $0.000023 =$
	3) $52,000,000 =$	4) $21,000 =$
	5) $3 \times 10^{-1} =$	6) $5 \times 10^{-2} =$
	7) $1.2 \times 10^3 =$	8) $2 \times 10^{-4} =$

Name:	Date: ....

Topic	**Scientific Notation - Answers**
Notes	✓ It is used to write very big or very small numbers in decimal form. ✓ In scientific notation all numbers are written in the form of: $$m \times 10^n$$ **Decimal notation** **Scientific notation** 3 3×10^0 $-45,000$ -4.5×10^4 0.3 3×10^{-1} $2,122.456$ 2.122456×10^3
Example	**Write 0.00054 in scientific notation.** First, move the decimal point to the right so that you have a number that is between 1 and 10. Then: $m = 5.4$ Now, determine how many places the decimal moved in step 1 by the power of 10. Then: $10^{-4} \rightarrow$ When the decimal moved to the right, the exponent is negative. Then: $0.00054 = 5.4 \times 10^{-4}$

Your Turn!	1) $0.000325 = 3.25 \times 10^{-4}$	2) $0.00023 = 2.3 \times 10^{-5}$
	3) $52,000,000 = 5.2 \times 10^7$	4) $21,000 = 2.1 \times 10^4$
	5) $3 \times 10^{-1} = 0.3$	6) $5 \times 10^{-2} = 0.05$
	7) $1.2 \times 10^3 = 1,200$	8) $2 \times 10^{-4} = 0.0002$

Name:	Date: ....

Topic	**Simplifying Polynomials**	
Notes	✓ Find "like" terms. (they have same variables with same power). ✓ Use "FOIL". (First–Out–In–Last) for binomials: $$(x + a)(x + b) = x^2 + (b + a)x + ab$$ ✓ Add or Subtract "like" terms using order of operation.	
Example	**Simplify this expression.** $(x + 3)(x - 8) =$ **Solution:** First apply FOIL method: $(a + b)(c + d) = ac + ad + bc + bd$ $(x + 3)(x - 8) = x^2 - 8x + 3x - 24$ Now combine like terms: $x^2 - 8x + 3x - 24 = x^2 - 5x - 24$	
Your Turn!	1) $-(2x - 4) =$	2) $2(2x + 6) =$
	3) $3x(3x - 4) =$	4) $5x(2x + 8) =$
	5) $-2x(5x + 6) + 5x =$	6) $-4x(8x - 3) - x^2 =$
	7) $(x + 4)(x + 5) =$	8) $(x + 2)(x + 8) =$
	9) $-4x^2 + 10x^3 + 5x^2 =$	10) $-3x^5 + 10x^4 + 5x^5 =$

| Name: .... | Date: .... |

Topic	**Simplifying Polynomials - Answers**
Notes	✓ Find "like" terms. (they have same variables with same power). ✓ Use "FOIL". (First–Out–In–Last) for binomials: $$(x + a)(x + b) = x^2 + (b + a)x + ab$$ ✓ Add or Subtract "like" terms using order of operation.
Example	**Simplify this expression.** $(x + 3)(x - 8) =$ **Solution:** First apply FOIL method: $(a + b)(c + d) = ac + ad + bc + bd$ $(x + 3)(x - 8) = x^2 - 8x + 3x - 24$ Now combine like terms: $x^2 - 8x + 3x - 24 = x^2 - 5x - 24$

Your Turn!	1) $-(2x - 4) =$
	   $-2x + 4$
		2) $2(2x + 6) =$
		   $4x + 12$

	3) $3x(3x - 4) =$
	   $9x^2 - 12x$
		4) $5x(2x + 8) =$
		   $10x^2 + 40x$

	5) $-2x(5x + 6) + 5x =$
	   $-10x^2 - 7x$
		6) $-4x(8x - 3) - x^2 =$
		   $-33x^2 + 12x$

	7) $(x + 4)(x + 5) =$
	   $x^2 + 9x + 20$
		8) $(x + 2)(x + 8) =$
		   $x^2 + 10x + 16$

	9) $-4x^2 + 10x^3 + 5x^2 =$
	   $10x^3 + x^2$
		10) $-3x^5 + 10x^4 + 5x^5 =$
		   $2x^5 + 10x^4$

Name:	Date: ....

Topic	**Adding and Subtracting Polynomials**	
Notes	✓ Adding polynomials is just a matter of combining like terms, with some order of operations considerations thrown in. ✓ Be careful with the minus signs, and don't confuse addition and multiplication!	
Example	**Simplify the expressions.** $(3x^2 - 4x^3) - (5x^3 - 8x^2) =$ **Solution:** First use Distributive Property: $-(5x^3 - 8x^2) = -5x^3 + 8x^2$ $\rightarrow (3x^2 - 4x^3) - (5x^3 - 8x^2) = 3x^2 - 4x^3 - 5x^3 + 8x^2$ Now combine like terms: $3x^2 - 4x^3 - 5x^3 + 8x^2 = -9x^3 + 11x^2$	
Your Turn!	1) $(x^2 - x) + (4x^2 - 5) =$ _____	2) $(2x^3 + x) - (x^3 + 2) =$ _____
	3) $(x^2 - 5x) + (6x^2 - 5) =$ _____	4) $(8x^2 - 2) - (3x^2 + 7) =$ _____
	5) $(3x^2 + 2) - (2 - 4x^2) =$ _____	6) $(x^3 + x^2) - (x^3 - 10) =$ _____
	7) $(3x^3 - 2x) - (x - x^3) =$ _____	8) $(x - 5x^4) - (2x^4 + 3x) =$ _____
	9) $(6x^3 + 5) - (4 - 5x^3) =$ _____	10) $(2x^2 + 5x^3) - (6x^3 + 7) =$ _____

| Name: .... | Date: .... |

Topic	**Adding and Subtracting Polynomials - Answers**
Notes	✓ Adding polynomials is just a matter of combining like terms, with some order of operations considerations thrown in. ✓ Be careful with the minus signs, and don't confuse addition and multiplication!
Example	**Simplify the expressions.** $(3x^2 - 4x^3) - (5x^3 - 8x^2) =$ **Solution:** First use Distributive Property: $-(5x^3 - 8x^2) = -5x^3 + 8x^2$ $\rightarrow (3x^2 - 4x^3) - (5x^3 - 8x^2) = 3x^2 - 4x^3 - 5x^3 + 8x^2$ Now combine like terms: $3x^2 - 4x^3 - 5x^3 + 8x^2 = -9x^3 + 11x^2$

Your Turn!

1) $(x^2 - x) + (4x^2 - 5) =$
$5x^2 - x - 5$

2) $(2x^3 + x) - (x^3 + 2) =$
$x^3 + x - 2$

3) $(x^2 - 5x) + (6x^2 - 5) =$
$7x^2 - 5x - 5$

4) $(8x^2 - 2) - (3x^2 + 7) =$
$5x^2 - 9$

5) $(3x^2 + 2) - (2 - 4x^2) =$
$7x^2$

6) $(x^3 + x^2) - (x^3 - 10) =$
$x^2 + 10$

7) $(3x^3 - 2x) - (x - x^3) =$
$4x^3 - 3x$

8) $(x - 5x^4) - (2x^4 + 3x) =$
$-7x^4 - 2x$

9) $(6x^3 + 5) - (4 - 5x^3) =$
$11x^3 + 1$

10) $(2x^2 + 5x^3) - (6x^3 + 7) =$
$-x^3 + 2x^2 - 7$

Name:	Date: ....

Topic	**Multiplying Binomials**
Notes	✓A binomial is a polynomial that is the sum or the difference of two terms, each of which is a monomial. ✓To multiply two binomials, use "FOIL" method. (First–Out–In–Last) $(x + a)(x + b) = x \times x + x \times b + a \times x + a \times b = x^2 + bx + ax + ab$
Example	**Multiply.** $(x - 4)(x + 9) =$ **Solution:** Use "FOIL". (First–Out–In–Last): $(x - 4)(x + 9) = x^2 + 9x - 4x - 36$ Then simplify: $x^2 + 9x - 4x - 36 = x^2 + 5x - 36$
Your Turn!	1) $(x + 2)(x + 2) =$ _____ 2) $(x + 3)(x + 2) =$ _____ 3) $(x - 3)(x + 4) =$ _____ 4) $(x - 2)(x - 4) =$ _____ 5) $(x + 3)(x + 4) =$ _____ 6) $(x + 5)(x + 4) =$ _____ 7) $(x - 6)(x - 5) =$ _____ 8) $(x - 5)(x - 5) =$ _____ 9) $(x + 6)(x - 8) =$ _____ 10) $(x - 9)(x + 7) =$ _____

Name:	Date: ....

Topic	**Multiplying Binomials - Answers**
Notes	✓A binomial is a polynomial that is the sum or the difference of two terms, each of which is a monomial. ✓To multiply two binomials, use "FOIL" method. (First–Out–In–Last) $(x + a)(x + b) = x \times x + x \times b + a \times x + a \times b = x^2 + bx + ax + ab$
Example	**Multiply.** $(x - 4)(x + 9) =$ **Solution:** Use "FOIL". (First–Out–In–Last): $(x - 4)(x + 9) = x^2 + 9x - 4x - 36$ Then simplify: $x^2 + 9x - 4x - 36 = x^2 + 5x - 36$

Your Turn!	1) $(x + 2)(x + 2) =$ $x^2 + 4x + 4$	2) $(x + 3)(x + 2) =$ $x^2 + 5x + 6$
	3) $(x - 3)(x + 4) =$ $x^2 + x - 12$	4) $(x - 2)(x - 4) =$ $x^2 - 6x + 8$
	5) $(x + 3)(x + 4) =$ $x^2 + 7x + 12$	6) $(x + 5)(x + 4) =$ $x^2 + 9x + 20$
	7) $(x - 6)(x - 5) =$ $x^2 - 11x + 30$	8) $(x - 5)(x - 5) =$ $x^2 - 10x + 25$
	9) $(x + 6)(x - 8) =$ $x^2 - 2x - 48$	10) $(x - 9)(x + 7) =$ $x^2 - 2x - 63$

| Name: .... | Date: .... |

Topic	**Multiplying and Dividing Monomials**
Notes	✓ When you divide or multiply two monomials you need to divide or multiply their coefficients and then divide or multiply their variables. ✓ In case of exponents with the same base, you need to subtract their powers. ✓ Exponent's rules: $$x^a \times x^b = x^{a+b}, \qquad \frac{x^a}{x^b} = x^{a-b}$$ $$\frac{1}{x^b} = x^{-b}, \quad (x^a)^b = x^{a \times b}$$ $$(xy)^a = x^a \times y^a$$
Example	**Divide expressions.** $\frac{-18x^5y^6}{2xy^2} =$ **Solution:** Use exponents' division rule: $\frac{x^a}{x^b} = x^{a-b}$, $\frac{x^5}{x} = x^{5-1} = x^4$ and $\frac{y^6}{y^2} = y^4$ Then: $\frac{-18x^5y^6}{2xy^2} = -9x^4y^4$
Your Turn!	1) $(x^8y)(xy^2) =$ _____	2) $(x^4y^3)(x^2y^3) =$ _____
	3) $(x^7y^4)(2x^5y^2) =$ _____	4) $(3x^5y^4)(4x^6y^3) =$ _____
	5) $(-6x^8y^7)(4x^6y^9) =$ _____	6) $(-2x^9y^3)(9x^7y^8) =$ _____
	7) $\frac{30x^8y^9}{6x^5y^4} =$ _____	8) $\frac{-42x^{12}y^{16}}{7x^8y^9} =$ _____

Name:	Date: ....

Topic	**Multiplying and Dividing Monomials - Answers**
Notes	✓ When you divide or multiply two monomials you need to divide or multiply their coefficients and then divide or multiply their variables. ✓ In case of exponents with the same base, you need to subtract their powers. ✓ Exponent's rules: $$x^a \times x^b = x^{a+b}, \qquad \frac{x^a}{x^b} = x^{a-b}$$ $$\frac{1}{x^b} = x^{-b}, \qquad (x^a)^b = x^{a \times b}$$ $$(xy)^a = x^a \times y^a$$
Example	**Divide expressions.** $\dfrac{-18x^5y^6}{2xy^2} =$ **Solution:** Use exponents' division rule: $\dfrac{x^a}{x^b} = x^{a-b}, \dfrac{x^5}{x} = x^{5-1} = x^4$ and $\dfrac{y^6}{y^2} = y^4$ Then: $\dfrac{-18x^5y^6}{2xy^2} = -9x^4y^4$
Your Turn!	1) $(x^8y)(xy^2) =$ x^9y^3 2) $(x^4y^3)(x^2y^3) =$ x^6y^6 3) $(x^7y^4)(2x^5y^2) =$ $2x^{12}y^6$ 4) $(3x^5y^4)(4x^6y^3) =$ $12x^{11}y^7$ 5) $(-6x^8y^7)(4x^6y^9) =$ $-24x^{14}y^{16}$ 6) $(-2x^9y^3)(9x^7y^8) =$ $-18x^{16}y^{11}$ 7) $\dfrac{30x^8y^9}{6x^5y^4} =$ $5x^3y^5$ 8) $\dfrac{-42x^{12}y^{16}}{7x^8y^9} =$ $-6x^4y^7$

Name:	Date: ....

Topic	**Multiplying a Polynomial and a Monomial**
Notes	✓ When multiplying monomials, use the product rule for exponents. $$x^a \times x^b = x^{a+b}$$ ✓ When multiplying a monomial by a polynomial, use the distributive property. $$a \times (b + c) = a \times b + a \times c = ab + ac$$ $$a \times (b - c) = a \times b - a \times c = ab - ac$$
Example	**Multiply expressions.** $4x(5x - 8) =$ **Solution:** Use Distributive Property: $4x(5x - 8) = 4x \times 5x - 4x \times (8) =$ Now, simplify: $4x \times 5x - 4x \times (8) = 20x^2 - 32x$
Your Turn!	1) $3x(2x + y) =$ _____ 2) $x(x - 3y) =$ _____ 3) $-x(5x - 3y) =$ _____ 4) $4x(x + 5y) =$ _____ 5) $-x(5x + 8y) =$ _____ 6) $2x(6x - 7y) =$ _____ 7) $-3x(x^3 + 4y^2 - 6x) =$ _____ 8) $7x(x^2 - 5y^2 + 4) =$ _____

Name:	Date: ....

Topic	**Multiplying a Polynomial and a Monomial - Answers**
Notes	✓ When multiplying monomials, use the product rule for exponents. $$x^a \times x^b = x^{a+b}$$ ✓ When multiplying a monomial by a polynomial, use the distributive property. $$a \times (b + c) = a \times b + a \times c = ab + ac$$ $$a \times (b - c) = a \times b - a \times c = ab - ac$$
Example	**Multiply expressions.** $4x(5x - 8) =$ **Solution:** Use Distributive Property: $4x(5x - 8) = 4x \times 5x - 4x \times (8) =$ Now, simplify: $4x \times 5x - 4x \times (8) = 20x^2 - 32x$

Your Turn!

1) $3x(2x + y) =$

$6x^2 + 3xy$

2) $x(x - 3y) =$
$x^2 - 3xy$

3) $-x(5x - 3y) =$

$-5x^2 + 3xy$

4) $4x(x + 5y) =$
$4x^2 + 20xy$

5) $-x(5x + 8y) =$

$-5x^2 - 8xy$

6) $2x(6x - 7y) =$
$12x^2 - 14xy$

7) $-3x(x^3 + 4y^2 - 6x) =$

$-3x^4 - 12xy^2 + 18x^2$

8) $7x(x^2 - 5y^2 + 4) =$
$7x^3 - 35xy^2 + 28x$

| Name: .... | Date: .... |

Topic	**Multiplying Monomials**
Notes	✓ A monomial is a polynomial with just one term: Examples: $5x$ or $7x^2yz^8$. ✓ When you multiply monomials, first multiply the coefficients (a number placed before and multiplying the variable) and then multiply the variables using multiplication property of exponents. $x^a \times x^b = x^{a+b}$
Example	**Multiply.** $(-3xy^4z^5) \times (2x^2y^5z^3) =$ **Solution:** Multiply coefficients and find same variables and use multiplication property of exponents: $x^a \times x^b = x^{a+b}$ $-3 \times 2 = -6$, $x \times x^2 = x^{1+2} = x^3$, $y^4 \times y^5 = y^{4+5} = y^9$, and $z^5 \times z^3 = z^{5+3} = z^8$ Then: $(-3xy^4z^5) \times (2x^2y^5z^3) = -6x^3y^9z^8$
Your Turn!	1) $2x^2 \times 4x^6 =$ _____ 2) $5x^7 \times 6x^4 =$ _____ 3) $-2x^2y^4 \times 6x^3y^2 =$ _____ 4) $-5x^5y \times 3x^3y^4 =$ _____ 5) $8x^7y^5 \times 5x^6y^3 =$ _____ 6) $-6x^7y^5 \times (-3x^9y^8) =$ _____ 7) $12x^8y^8z^4 \times 3x^4y^3z =$ _____ 8) $-8x^9y^7z^{11} \times 7x^6y^7z^5 =$ _____

Name:	Date: ....

Topic	**Multiplying Monomials - Answers**
Notes	✓ A monomial is a polynomial with just one term: Examples: $5x$ or $7x^2yz^8$. ✓ When you multiply monomials, first multiply the coefficients (a number placed before and multiplying the variable) and then multiply the variables using multiplication property of exponents. $x^a \times x^b = x^{a+b}$
Example	**Multiply.** $(-3xy^4z^5) \times (2x^2y^5z^3) =$ **Solution:** Multiply coefficients and find same variables and use multiplication property of exponents: $x^a \times x^b = x^{a+b}$ $-3 \times 2 = -6, x \times x^2 = x^{1+2} = x^3$, $y^4 \times y^5 = y^{4+5} = y^9$, and $z^5 \times z^3 = z^{5+3} = z^8$ Then: $(-3xy^4z^5) \times (2x^2y^5z^3) = -6x^3y^9z^8$

Your Turn!	1) $2x^2 \times 4x^6 =$ $8x^8$	2) $5x^7 \times 6x^4 =$ $30x^{11}$
	3) $-2x^2y^4 \times 6x^3y^2 =$ $-12x^5y^6$	4) $-5x^5y \times 3x^3y^4 =$ $-15x^8y^5$
	5) $8x^7y^5 \times 5x^6y^3 =$ $40x^{13}y^8$	6) $-6x^7y^5 \times (-3x^9y^8) =$ $18x^{16}y^{13}$
	7) $12x^8y^8z^4 \times 3x^4y^3z =$ $36x^{12}y^{11}z^5$	8) $-8x^9y^7z^{11} \times 7x^6y^7z^5 =$ $-56x^{15}y^{14}z^{16}$

Name:	Date: ....

Topic	Factoring Trinomials
Notes	To factor trinomial, use of the following methods: ✓ "FOIL": $(x + a)(x + b) = x^2 + (b + a)x + ab$ ✓ "Difference of Squares": $$a^2 - b^2 = (a + b)(a - b)$$ $$a^2 + 2ab + b^2 = (a + b)(a + b)$$ $$a^2 - 2ab + b^2 = (a - b)(a - b)$$ ✓ "Reverse FOIL": $x^2 + (b + a)x + ab = (x + a)(x + b)$
Example	**Factor this trinomial.** $x^2 + 12x + 32 =$ **Solution:** Break the expression into groups: $(x^2 + 4x) + (8x + 32)$ Now factor out x from $x^2 + 4x : x(x + 4)$, and factor out 8 from $8x + 32$: $8(x + 4)$ Then: $(x^2 + 4x) + (8x + 32) = x(x + 4) + 8(x + 4)$ Now factor out like term: $(x + 4) \rightarrow (x + 4)(x + 8)$
Your Turn!	1) $x^2 + 6x + 9 =$ ——— ⎮ 2) $x^2 + 5x + 6 =$ ——— 3) $x^2 + x - 12 =$ ——— ⎮ 4) $x^2 - 6x + 8 =$ ——— 5) $x^2 + 7x + 12 =$ ——— ⎮ 6) $x^2 + 12x + 32 =$ ——— 7) $x^2 - 11x + 30 =$ ——— ⎮ 8) $x^2 - 14x + 45 =$ ———

Name: ....   **Date:** ....

Topic	Factoring Trinomials - Answers
Notes	To factor trinomial, use of the following methods: ✓ "FOIL": $(x + a)(x + b) = x^2 + (b + a)x + ab$ ✓ "Difference of Squares": $$a^2 - b^2 = (a + b)(a - b)$$ $$a^2 + 2ab + b^2 = (a + b)(a + b)$$ $$a^2 - 2ab + b^2 = (a - b)(a - b)$$ ✓ "Reverse FOIL": $x^2 + (b + a)x + ab = (x + a)(x + b)$

Example	**Factor this trinomial.** $x^2 + 12x + 32 =$
	   Solution: Break the expression into groups: $(x^2 + 4x) + (8x + 32)$
	   Now factor out x from $x^2 + 4x$: $x(x + 4)$, and factor out 8 from $8x + 32$: $8(x + 4)$
	   Then: $(x^2 + 4x) + (8x + 32) = x(x + 4) + 8(x + 4)$
	   Now factor out like term: $(x + 4) \rightarrow (x + 4)(x + 8)$

Your Turn!	1) $x^2 + 6x + 9 =$ $(x + 3)(x + 3)$	2) $x^2 + 5x + 6 =$ $(x + 3)(x + 2)$
	3) $x^2 + x - 12 =$ $(x - 3)(x + 4)$	4) $x^2 - 6x + 8 =$ $(x - 2)(x - 4)$
	5) $x^2 + 7x + 12 =$ $(x + 3)(x + 4)$	6) $x^2 + 12x + 32 =$ $(x + 8)(x + 4)$
	7) $x^2 - 11x + 30 =$ $(x - 6)(x - 5)$	8) $x^2 - 14x + 45 =$ $(x - 9)(x - 5)$

Name:	Date: ....

Topic	The Pythagorean Theorem
Notes	✓ In any right triangle: $a^2 + b^2 = c^2$
Example	**Right triangle ABC (not shown) has two legs of lengths 18 cm (AB) and 24 cm (AC). What is the length of the third side (BC)?** **Solution:** Use Pythagorean Theorem: $a^2 + b^2 = c^2$ Then: $a^2 + b^2 = c^2 \rightarrow 18^2 + 24^2 = c^2 \rightarrow 324 + 576 = c^2$ $c^2 = 900 \rightarrow c = \sqrt{900} = 30\ cm$
Your Turn!	1) _____ 15 ? 8 2) _____ 34 16 ? 3) _____ 13 5 ? 4) _____ 15 ? 12

Name:	Date: ....

Topic	The Pythagorean Theorem - Answers
Notes	✓ In any right triangle: $a^2 + b^2 = c^2$
Example	**Right triangle ABC (not shown) has two legs of lengths 18 cm (AB) and 24 cm (AC). What is the length of the third side (BC)?** **Solution:** Use Pythagorean Theorem: $a^2 + b^2 = c^2$ Then: $a^2 + b^2 = c^2 \rightarrow 18^2 + 24^2 = c^2 \rightarrow 324 + 576 = c^2$ $c^2 = 900 \rightarrow c = \sqrt{900} = 30 \; cm$
Your Turn!	1) 17 15 ? 8 2) 30 16 34 ? 3) 12 13 5 ? 4) 9 15 ? 12

Name:	Date: ....

Topic	**Parallel lines and Transversals**

Notes	✓ When a line (transversal) intersects two parallel lines in the same plane, eight angles are formed. In the following diagram, a transversal intersects two parallel lines. Angles 1, 7, 3, and 5 are congruent. Angles 2, 8, 4, and 6 are also congruent.
	✓ In the following diagram, the following angles are supplementary angles (their sum is 180):
	  - Angles 1 and 8
	  - Angles 2 and 7
	  - Angles 3 and 6
	  - Angles 4 and 5

Example	**In the following diagram, two parallel lines are cut by a transversal. What is the value of x?** **Solution:** The two angles 75° and $11x - 2$ are equal. $11x - 2 = 75$ Now, solve for x: $11x - 2 + 2 = 75 + 2 \rightarrow$ $11x = 77 \rightarrow x = \dfrac{77}{11} \rightarrow x = 7$
Your Turn!	1) Find the measure of the angle indicated. $? =$ ___ $100°$ $?$ 2) Solve for x. $x =$ ___ $x + 139$ $132°$

Name:	Date: ....

Topic	**Parallel lines and Transversals - Answers**

Notes	✓ When a line (transversal) intersects two parallel lines in the same plane, eight angles are formed. In the following diagram, a transversal intersects two parallel lines. Angles 1, 7, 3, and 5 are congruent. Angles 2, 8, 4, and 6 are also congruent.
	✓ In the following diagram, the following angles are supplementary angles (their sum is 180):
	  - Angles 1 and 8
	  - Angles 2 and 7
	  - Angles 3 and 6
	  - Angles 4 and 5

Example	**In the following diagram, two parallel lines are cut by a transversal. What is the value of x?** **Solution:** The two angles 75° and $11x - 2$ are equal. $11x - 2 = 75$ Now, solve for x: $11x - 2 + 2 = 75 + 2 \rightarrow$ $11x = 77 \rightarrow x = \dfrac{77}{11} \rightarrow x = 7$
Your Turn!	1) Find the measure of the angle indicated. $? = 80°$ 100° ? 2) Solve for x. $x = -7$ $x + 139$ 132°

Name:	Date: ....

Topic	**Triangles**
Notes	✓ In any triangle the sum of all angles is 180 degrees. ✓ Area of a triangle = $\frac{1}{2}(base \times height)$
Example	**What is the area of the following triangle?** **Solution:** Use the area formula: Area $= \frac{1}{2}(base \times height)$ $base = 16$ and $height = 6$ Area $= \frac{1}{2}(16 \times 6) = \frac{96}{2} = 48$
Your Turn!	1) _____ 2) _____ 3) _____ 4) _____

Name:	Date: ....

Topic	**Triangles - Answers**
Notes	✓ In any triangle the sum of all angles is 180 degrees. ✓ Area of a triangle = $\frac{1}{2}(base \times height)$
Example	**What is the area of the following triangle?** **Solution:** Use the area formula: Area $= \frac{1}{2}(base \times height)$ $base = 16$ and $height = 6$ Area $= \frac{1}{2}(16 \times 6) = \frac{96}{2} = 48$
Your Turn!	1) 120 2) 252 3) 300 4) 736

Name: ....   **Date:** ....

Topic	Special Right Triangles
Notes	✓ A special right triangle is a triangle whose sides are in a particular ratio. Two special right triangles are $45° - 45° - 90°$ and $30° - 60° - 90°$ triangles. ✓ In a special $45° - 45° - 90°$ triangle, the three angles are $45°, 45°$ and $90°$. The lengths of the sides of this triangle are in the ratio of $1:1:\sqrt{2}$. ✓ In a special triangle $30° - 60° - 90°$, the three angles are $30° - 60° - 90°$. The lengths of this triangle are in the ratio of $1:\sqrt{3}:2$. *(triangle diagram: $30°$ top, sides $a\sqrt{3}$ and $2a$, $60°$ at bottom right, base a)*
Example	**Find the length of the hypotenuse of a right triangle if the length of the other two sides are both 5 inches.** *Solution:* this is a right triangle with two equal sides. Therefore, it must be a $45° - 45° - 90°$ triangle. Two equal sides are 5 inches. So, the length of the hypotenuse is $5\sqrt{2}$ inches. If the first and second value of the ratio $x:x:x\sqrt{2}$. $$x:x:x\sqrt{2} \to x = 5 \to 5:5:5\sqrt{2}$$
Your Turn!	**Find the value of x and y in each triangle.** 1) $x =$ ___ $y =$ ___ *(triangle with side x, 12, 30° angle, base y)* 2) $x =$ ___ $y =$ ___ *(triangle with side 7, y, 45° angle, base X)*

Name:	**Date:** ....

Topic	**Special Right Triangles - Answers**
Notes	✓ A special right triangle is a triangle whose sides are in a particular ratio. Two special right triangles are $45° - 45° - 90°$ and $30° - 60° - 90°$ triangles. ✓ In a special $45° - 45° - 90°$ triangle, the three angles are $45°$, $45°$ and $90°$. The lengths of the sides of this triangle are in the ratio of $1 : 1 : \sqrt{2}$. ✓ In a special triangle $30° - 60° - 90°$, the three angles are $30° - 60° - 90°$. The lengths of this triangle are in the ratio of $1 : \sqrt{3} : 2$.
Example	**Find the length of the hypotenuse of a right triangle if the length of the other two sides are both 5 inches.** *Solution:* this is a right triangle with two equal sides. Therefore, it must be a $45° - 45° - 90°$ triangle. Two equal sides are 5 inches. So, the length of the hypotenuse is $5\sqrt{2}$ inches. If the first and second value of the ratio $x : x : x\sqrt{2}$. $$x : x : x\sqrt{2} \rightarrow x = 5 \rightarrow 5 : 5 : 5\sqrt{2}$$
Your Turn!	**Find the value of x and y in each triangle.** 1) $x = 24 \quad y = 12\sqrt{3}$ 2) $x = 7 \quad y = 7\sqrt{2}$

Name: ....

Date: ....

Topic	Polygons
Notes	Perimeter of a square $= 4 \times side = 4s$ Perimeter of a rectangle $= 2(width + length)$ Perimeter of a trapezoid $= a + b + c + d$ Perimeter of a regular hexagon $= 6a$ Perimeter of a parallelogram $= 2(l + w)$
Example	**Find the perimeter of following regular hexagon.** **Solution:** Since the hexagon is regular, all sides are equal. Then: Perimeter of Hexagon $= 6 \times (one\ side)$ Perimeter of Hexagon $= 6 \times (one\ side) = 6 \times 9 = 54\ m$
Your Turn!	1) (rectangle) _____ $9\ in$ $15\ in$ 2) (trapezoid)_____ $8\ m$ $10\ m$ $10\ m$ $14\ m$ 3) (regular hexagon)_____ $5\ m$ 4) (parallelogram)_____ $10\ in$ $16\ in$

Name:	Date: ....

Topic	Polygons - Answers
Notes	Perimeter of a square $= 4 \times side = 4s$ Perimeter of a rectangle $= 2(width + length)$ Perimeter of a trapezoid $= a + b + c + d$ Perimeter of a regular hexagon $= 6a$ Perimeter of a parallelogram $= 2(l + w)$
Example	**Find the perimeter of following regular hexagon.** **Solution:** Since the hexagon is regular, all sides are equal. Then: Perimeter of Hexagon $= 6 \times (one\ side)$ Perimeter of Hexagon $= 6 \times (one\ side) = 6 \times 9 = 54\ m$

Your Turn!	1) (rectangle) *48 in* 9 *in* 15 *in*	2) *42 m* 8 *m* 10 *m* 10 *m* 14 *m*
	3) (regular hexagon) *30 m* 5 *m*	4) (parallelogram) *52 in* 10 *in* 16 *in*

Name:	Date: ....

Topic	**Cubes**
Notes	✓ A cube is a three-dimensional solid object bounded by six square sides. ✓ Volume is the measure of the amount of space inside of a solid figure, like a cube, ball, cylinder or pyramid. ✓ Volume of a cube $= (one\ side)^3$ ✓ surface area of cube $= 6 \times (one\ side)^2$

Example	**Find the volume and surface area of the following cube.**
	  Solution:
	Use volume formula: $volume = (one\ side)^3$
	Then: $volume = (one\ side)^3 = (15)^3 = 3,375\ cm^3$
	Use surface area formula:
	$surface\ area\ of\ cube: 6(one\ side)^2 = 6(15)^2 = 6(225) = 1,350\ cm^2$
	     15 cm

Your Turn!	**Find the volume of each cube.**
	
	1) _____
	    11 in
	
	2) _____
	    13 ft
	
	3) _____
	    14 cm
	
	4) _____
	    30 m

| Name: .... | Date: .... |

Topic	Cubes - Answers
Notes	✓ A cube is a three-dimensional solid object bounded by six square sides. ✓ Volume is the measure of the amount of space inside of a solid figure, like a cube, ball, cylinder or pyramid. ✓ Volume of a cube $= (one\ side)^3$ ✓ surface area of cube $= 6 \times (one\ side)^2$

Example

Find the volume and surface area of the following cube.

15 cm

Solution:

Use volume formula: $volume = (one\ side)^3$

Then: $volume = (one\ side)^3 = (15)^3 = 3,375\ cm^3$

Use surface area formula:

$surface\ area\ of\ cube: 6(one\ side)^2 = 6(15)^2 = 6(225) = 1,350\ cm^2$

Your Turn!

Find the volume of each cube.

1) $1,331\ in^3$

11 in

2) $2,197\ ft^3$

13 ft

3) $2,744\ cm^3$

14 cm

4) $27,000\ m^3$

30 m

Name:	Date: ....

Topic	**Trapezoids**
Notes	✓ A quadrilateral with at least one pair of parallel sides is a trapezoid. ✓ Area of a trapezoid $= \frac{1}{2}h(b_1 + b_2)$
Example	**Calculate the area of the trapezoid.** **Solution:** Use area formula: $A = \frac{1}{2}h(b_1 + b_2)$ $b_1 = 8\ cm$, $b_2 = 12\ cm$ and $h = 14\ cm$ Then: $A = \frac{1}{2}(14)(12 + 8) = 7(20) = 140\ cm^2$
Your Turn!	1) _____ *5 cm, 4 cm, 9 cm* 2) _____ *8 m, 10 m, 12 m* 3) _____ *7 ft, 6 ft, 15 ft* 4) _____ *10 cm, 8 cm, 14 cm*

Name:	Date: ....

Topic	**Trapezoids - Answers**
Notes	✓ A quadrilateral with at least one pair of parallel sides is a trapezoid. ✓ Area of a trapezoid $= \frac{1}{2}h(b_1 + b_2)$ b_2 h b_1
Example	Calculate the area of the trapezoid. **Solution:** Use area formula: $A = \frac{1}{2}h(b_1 + b_2)$ $b_1 = 8 \ cm$, $b_2 = 12 \ cm$ and $h = 14 \ cm$ Then: $A = \frac{1}{2}(14)(12 + 8) = 7(20) = 140 \ cm^2$ 12 cm 14 cm 8 cm

Your Turn!

1) $28 \ cm^2$

5 cm
4 cm
9 cm

2) $100 \ m^2$

8 m
10 m
12 m

3) $66 \ ft^2$

7 ft
6 ft
15 ft

4) $96 \ cm^2$

10 cm
8 cm
14 cm

Name:	Date: ....

Topic	Rectangular Prisms
Notes	✓ A solid 3-dimensional object which has six rectangular faces. ✓ Volume of a Rectangular prism $= Length \times Width \times Height$ $Volume = l \times w \times h$ $Surface\ area = 2(wh + lw + lh)$
Example	**Find the volume and surface area of rectangular prism.** **Solution:** Use volume formula: $Volume = l \times w \times h$ Then: $Volume = 4 \times 2 \times 6 = 48\ m^3$ Use surface area formula: $Surface\ area = 2(wh + lw + lh)$ Then: $Surface\ area = 2\big((2 \times 6) + (4 \times 2) + (4 \times 6)\big) =$ $2(12 + 8 + 24) = 2(44) = 88\ m^2$
Your Turn!	**Find the surface area of each Rectangular Prism.** 1) _____ 6 ft, 10 ft, 4 ft 2) _____ 8 cm, 16 cm, 6 cm 3) _____ 12 m, 18 m, 10 m 4) _____ 20 in, 15 in, 12 in

Name:	Date: ....

Topic	**Rectangular Prisms - Answers**
Notes	✓ A solid 3-dimensional object which has six rectangular faces. ✓ Volume of a Rectangular prism $= Length \times Width \times Height$ $Volume = l \times w \times h$ $Surface\ area = 2(wh + lw + lh)$
Example	**Find the volume and surface area of rectangular prism.** **Solution:** Use volume formula: $Volume = l \times w \times h$ Then: $Volume = 4 \times 2 \times 6 = 48\ m^3$ Use surface area formula: $Surface\ area = 2(wh + lw + lh)$ Then: $Surface\ area = 2\big((2 \times 6) + (4 \times 2) + (4 \times 6)\big) =$ $2(12 + 8 + 24) = 2(44) = 88\ m^2$

Find the surface area of each Rectangular Prism.

Your Turn!

1) $248\ ft^2$

6 ft
10 ft
4 ft

2) $544\ cm^2$

8 cm
16 cm
6 cm

3) $1,032\ m^2$

12 m
18 m
10 m

4) $1,440\ in^2$

20 in
15 in
12 in

Name:	Date: ....

Topic	**Cylinder**
Notes	✓ A cylinder is a solid geometric figure with straight parallel sides and a circular or oval cross section. ✓ *Volume of Cylinder Formula* $= \pi(radius)^2 \times height$ $\pi = 3.14$ ✓ *Surface area of a cylinder* $= 2\pi r^2 + 2\pi rh$
Example	**Find the volume and Surface area of the follow Cylinder.** **Solution:** Use volume formula: $Volume = \pi(radius)^2 \times height$ Then: $Volume = \pi(3)^2 \times 12 = 9\pi \times 12 = 108\pi$ $\pi = 3.14$ then: $Volume = 108\pi = 339.12\ cm^3$ Use surface area formula: $Surface\ area = 2\pi r^2 + 2\pi rh$ Then: $2\pi(3)^2 + 2\pi(3)(12) = 2\pi(9) + 2\pi(36) = 18\pi + 72\pi = 90\pi$ $\pi = 3.14$ Then: $Surface\ area = 90 \times 3.14 = 282.6\ cm^2$ *12 cm* *3 cm*
Your Turn!	**Find the volume of each Cylinder.** ($\pi = 3.14$) 1) _____ *10 in* *2 in* 2) _____ *14 m* *5 m* **Find the Surface area of each Cylinder.** ($\pi = 3.14$) 3) _____ *15 ft* *9 ft* 4) _____ *20 cm* *12 cm*

Name: ....

Date: ....

Topic	Cylinder - Answers
Notes	✓ A cylinder is a solid geometric figure with straight parallel sides and a circular or oval cross section. ✓ *Volume of Cylinder Formula* $= \pi(radius)^2 \times height \quad \pi = 3.14$ ✓ *Surface area of a cylinder* $= 2\pi r^2 + 2\pi rh$ *height* *radius*
Example	**Find the volume and Surface area of the follow Cylinder.** **Solution:** Use volume formula: $Volume = \pi(radius)^2 \times height$ Then: $Volume = \pi(3)^2 \times 12 = 9\pi \times 12 = 108\pi$ $\pi = 3.14$ then: $Volume = 108\pi = 339.12\ cm^3$ Use surface area formula: $Surface\ area = 2\pi r^2 + 2\pi rh$ Then: $2\pi(3)^2 + 2\pi(3)(12) = 2\pi(9) + 2\pi(36) = 18\pi + 72\pi = 90\pi$ $\pi = 3.14$ Then: $Surface\ area = 90 \times 3.14 = 282.6\ cm^2$ *12 cm* *3 cm*
Your Turn!	**Find the volume of each Cylinder.** $(\pi = 3.14)$ 1) $125.6\ in^3$ 2) $1,099\ m^3$ *10 in* *2 in* *14 m* *5 m* **Find the Surface area of each Cylinder.** $(\pi = 3.14)$ 3) $1,356.48\ ft^2$ 4) $2,411.52\ cm^2$ *15 ft* *9 ft* *20 cm* *12 cm*

Name:	Date: ....

Topic	Mean, Median, Mode, and Range of the Given Data
Notes	✓ Mean: $\frac{sum\ of\ the\ data}{total\ number\ of\ data\ entires}$ ✓ Mode: value in the list that appears most often. ✓ Median: is the middle number of a group of numbers that have been arranged in order by size. ✓ Range: the difference of largest value and smallest value in the list.
Example	**Find the mode and median of these numbers?** $\mathbf{16, 10, 6, 3, 1, 16, 2, 4}$ **Solution:** Mode: value in the list that appears most often. Number 16 is the value in the list that appears most often (there are two number 16). To find median, write the numbers in order: $1, 2, 3, 4, 6, 10, 16, 16$ Number 4 and 6 are in the middle. Find their average: $\frac{4+6}{2} = \frac{10}{2} = 5$ The median is 5.
Your Turn!	1) $3, 2, 4, 8, 3, 10$ Mode: _____ Range: _____ Mean: _____ Median: _____
	2) $6, 3, 2, 9, 5, 7, 2, 14$ Mode: _____ Range: _____ Mean: _____ Median: _____
	3) $5, 4, 3, 2, 9, 5, 6, 8, 12$ Mode: _____ Range: _____ Mean: _____ Median: _____
	4) $12, 6, 8, 6, 9, 6, 4, 13$ Mode: _____ Range: _____ Mean: _____ Median: _____

Name:	Date: ....

Topic	Mean, Median, Mode, and Range of the Given Data - Answers
Notes	✓ Mean: $\dfrac{sum\ of\ the\ data}{total\ number\ of\ data\ entires}$ ✓ Mode: value in the list that appears most often. ✓ Median: is the middle number of a group of numbers that have been arranged in order by size. ✓ Range: the difference of largest value and smallest value in the list.
Example	**Find the mode and median of these numbers? 16, 10, 6, 3, 1, 16, 2, 4** **Solution:** Mode: value in the list that appears most often. Number 16 is the value in the list that appears most often (there are two number 16). To find median, write the numbers in order: 1, 2, 3, 4, 6, 10, 16, 16 Number 4 and 6 are in the middle. Find their average: $\dfrac{4+6}{2} = \dfrac{10}{2} = 5$ The median is 5.

Your Turn!

1) 3, 2, 4, 8, 3, 10

     Mode: 3   Range: 8

     Mean: 5   Median: 3.5

2) 6, 3, 2, 9, 5, 7, 2, 14

     Mode: 2   Range: 12

     Mean: 6   Median: 5.5

3) 5, 4, 3, 2, 9, 5, 6, 8, 12

     Mode: 5   Range: 10

     Mean: 6   Median: 5

4) 12, 6, 8, 6, 9, 6, 4, 13

     Mode: 6   Range: 9

     Mean: 8   Median: 7

Name:	Date: ....

Topic	Probability Problems
Notes	✓ Probability is the likelihood of something happening in the future. It is expressed as a number between zero (can never happen) to 1 (will always happen). ✓ Probability can be expressed as a fraction, a decimal, or a percent. ✓ Probability formula: $Probability = \dfrac{number\ of\ desired\ outcomes}{number\ of\ total\ outcomes}$
Example	**If there are 3 green balls, 4 red balls, and 10 blue balls in a basket, what is the probability that Jason will pick out a red ball from the basket?** **Solution:** There are 4 red ball and 17 are total number of balls. Therefore, probability that Jason will pick out a red ball from the basket is 4 out of 17 or $\dfrac{4}{3+4+10} = \dfrac{4}{17}$
Your Turn!	1) A number is chosen at random from 1 to 20. Find the probability of selecting a prime number. (A prime number is a whole number that is only divisible by itself and 1) _____ 2) There are only red and blue cards in a box. The probability of choosing a red card in the box at random is one third. If there are 24 blue cards, how many cards are in the box? _____ 3) A die is rolled, what is the probability that an even number is obtained? _____

Name:	Date: ....

Topic	**Probability Problems - Answers**
Notes	✓ Probability is the likelihood of something happening in the future. It is expressed as a number between zero (can never happen) to 1 (will always happen). ✓ Probability can be expressed as a fraction, a decimal, or a percent. ✓ Probability formula: $Probability = \frac{number\ of\ desired\ outcomes}{number\ of\ total\ outcomes}$
Example	**If there are 3 green balls, 4 red balls, and 10 blue balls in a basket, what is the probability that Jason will pick out a red ball from the basket?** **Solution:** There are 4 red ball and 17 are total number of balls. Therefore, probability that Jason will pick out a red ball from the basket is 4 out of 17 or $\frac{4}{3+4+10} = \frac{4}{17}$
Your Turn!	1) A number is chosen at random from 1 to 20. Find the probability of selecting a prime number. (A prime number is a whole number that is only divisible by itself and 1) $\frac{8}{20} = \frac{2}{5}$ (There are 8 prime numbers from 1 to 20: $2, 3, 5, 7, 11, 13, 17, 19$) 2) There are only red and blue cards in a box. The probability of choosing a red card in the box at random is one third. If there are 24 blue cards, how many cards are in the box? 36 3) A die is rolled, what is the probability that an even number is obtained? $\frac{1}{2}$

Name:	Date: ....

Topic	**Pie Graph**
Notes	✓ A Pie Chart is a circle chart divided into sectors, each sector represents the relative size of each value.
Example	**A library has 460 books that include Mathematics, Physics, Chemistry, English and History. Use following graph to answer the question.** **What is the number of Physics books?** **Solution:** Number of total books $= 460$ Percent of Physics books $= 25\% = 0.25$ Then, umber of Physics books: $0.25 \times 460 = 115$ History 10% Mathematics 30% English 15% Chemistry 20% Physics 25%
Your Turn!	The circle graph below shows all Mr. Smith's expenses for last month. Mr. Smith spent $440 for clothes last month. Mr. Smith's last month expenses Foods 25% Bills 18% Others 23% Clothes 20% Books 14%
	1) How much did Mr. Smith spend for his Books last month? _____ 2) How much did Mr. Smith spend for Bills last month? _____ 3) How much did Mr. Smith spend for his foods last month? _____

Name: ....   **Date:** ....

Topic	Pie Graph - Answers
Notes	✓ A Pie Chart is a circle chart divided into sectors, each sector represents the relative size of each value.
Example	**A library has 460 books that include Mathematics, Physics, Chemistry, English and History. Use following graph to answer the question.** **What is the number of Physics books?** **Solution:** Number of total books $= 460$ Percent of Physics books $= 25\% = 0.25$ Then, umber of Physics books: $0.25 \times 460 = 115$

The circle graph below shows all Mr. Smith's expenses for last month. Mr. Smith spent $440 for clothes last month.

Mr. Smith's last month expenses

Your Turn!

1) How much did Mr. Smith spend for his Books last month? $308

2) How much did Mr. Smith spend for Bills last month? $396

3) How much did Mr. Smith spend for his foods last month? $550

Name:	Date: ....

Topic	**Permutations and Combinations**
Notes	✓ Permutations: The number of ways to choose a sample of k elements from a set of n distinct objects where order does matter, and replacements are not allowed. For a permutation problem, use this formula: $$_nP_k = \frac{n!}{(n-k)!}$$ ✓ Combination: The number of ways to choose a sample of r elements from a set of n distinct objects where order does not matter, and replacements are not allowed. For a combination problem, use this formula: $$_nC_r = \frac{n!}{r!\,(n-r)!}$$ ✓ Factorials are products, indicated by an exclamation mark. For example, 4! Equals: $4 \times 3 \times 2 \times 1$. Remember that 0! is defined to be equal to 1.
Example	**How many ways can we pick a team of 4 people from a group of 8?** **Solution:** Since the order doesn't matter, we need to use combination formula where n is 8 and r is 4. Then: $\frac{n!}{r!(n-r)!} = \frac{8!}{4!(8-4)!} = \frac{8!}{4!(4)!} = \frac{8\times7\times6\times5\times4!}{4!(4)!} = \frac{8\times7\times6\times5}{4\times3\times2\times1} = \frac{1,680}{24} = 70$
Your Turn!	1) In how many ways can 8 athletes be arranged in a straight line? _____ 2) How many ways can we award a first and second place prize among eight contestants? _____ 3) In how many ways can we choose 3 players from a team of 9 players? _____

| Name: .... | Date: .... |

Topic	**Permutations and Combinations - Answers**
Notes	✓ Permutations: The number of ways to choose a sample of k elements from a set of n distinct objects where order does matter, and replacements are not allowed. For a permutation problem, use this formula: $$_nP_k = \frac{n!}{(n-k)!}$$ ✓ Combination: The number of ways to choose a sample of r elements from a set of n distinct objects where order does not matter, and replacements are not allowed. For a combination problem, use this formula: $$_nC_r = \frac{n!}{r!\,(n-r)!}$$ ✓ Factorials are products, indicated by an exclamation mark. For example, 4! Equals: $4 \times 3 \times 2 \times 1$. Remember that 0! is defined to be equal to 1.
Example	**How many ways can we pick a team of 4 people from a group of 8?** **Solution:** Since the order doesn't matter, we need to use combination formula where n is 8 and r is 4. Then: $\frac{n!}{r!(n-r)!} = \frac{8!}{4!(8-4)!} = \frac{8!}{4!(4)!} = \frac{8\times7\times6\times5\times4!}{4!(4)!} = \frac{8\times7\times6\times5}{4\times3\times2\times1} = \frac{1,680}{24} = 70$
Your Turn!	1) In how many ways can 8 athletes be arranged in a straight line? 40,320
	2) How many ways can we award a first and second place prize among eight contestants? 56
	3) In how many ways can we choose 3 players from a team of 9 players? 84

Name:	Date: ....

Topic	Solving a Quadratic Equation
Notes	✓ Write the equation in the form of: $ax^2 + bx + c = 0$ ✓ Factor the quadratic and solve for the variable. ✓ Use quadratic formula if you couldn't factorize the quadratic. ✓ Quadratic formula: $x = \dfrac{-b \pm \sqrt{b^2 - 4ac}}{2a}$
Example	**Find the solutions of quadratic. $x^2 + x - 72 = 0$** **Solution:** Use quadratic formula: $x = \dfrac{-b \pm \sqrt{b^2 - 4ac}}{2a}$, $a = 1, b = 1$ and $c = -72$ $x = \dfrac{-1 \pm \sqrt{1^2 - 4 \times 1(-72)}}{2 \times 1}$ $x_1 = \dfrac{-1 + \sqrt{1^2 - 4 \times 1 \times (-72)}}{2 \times 1} = 8$, $x_2 = \dfrac{-1 - \sqrt{1^2 - 4 \times 1 \times (-72)}}{2 \times 1} = -9$

Your Turn!	1) $x^2 - x - 2 = 0$ $x = \underline{\quad}, x = \underline{\quad}$	2) $x^2 - 6x + 8 = 0$ $x = \underline{\quad}, x = \underline{\quad}$
	3) $x^2 - 4x + 3 = 0$ $x = \underline{\quad}, x = \underline{\quad}$	4) $x^2 + x - 12 = 0$ $x = \underline{\quad}, x = \underline{\quad}$
	5) $x^2 + 7x - 18 = 0$ $x = \underline{\quad}, x = \underline{\quad}$	6) $x^2 - 2x - 15 = 0$ $x = \underline{\quad}, x = \underline{\quad}$
	7) $x^2 + 6x - 40 = 0$ $x = \underline{\quad}, x = \underline{\quad}$	8) $x^2 - 9x - 36 = 0$ $x = \underline{\quad}, x = \underline{\quad}$

Name: ....

Date: ....

Topic	Solving a Quadratic Equation- Answers
Notes	✓ Write the equation in the form of: $ax^2 + bx + c = 0$ ✓ Factor the quadratic and solve for the variable. ✓ Use quadratic formula if you couldn't factorize the quadratic. ✓ Quadratic formula: $x = \frac{-b \pm \sqrt{b^2 - 4ac}}{2a}$
Example	**Find the solutions of quadratic.** $x^2 + x - 72 = 0$ **Solution:** Use quadratic formula: $x = \frac{-b \pm \sqrt{b^2 - 4ac}}{2a}$, $a = 1, b = 1$ and $c = -72$ $x = \frac{-1 \pm \sqrt{1^2 - 4 \times 1(-72)}}{2 \times 1}$ $x_1 = \frac{-1 + \sqrt{1^2 - 4 \times 1 \times (-72)}}{2 \times 1} = 8$, $x_2 = \frac{-1 - \sqrt{1^2 - 4 \times 1 \times (-72)}}{2 \times 1} = -9$
Your Turn!	1) $x^2 - x - 2 = 0$ $x = 2, x = -1$ 2) $x^2 - 6x + 8 = 0$ $x = 2, x = 4$ 3) $x^2 - 4x + 3 = 0$ $x = 3, x = 1$ 4) $x^2 + x - 12 = 0$ $x = 3, x = -4$ 5) $x^2 + 7x - 18 = 0$ $x = 2, x = -9$ 6) $x^2 - 2x - 15 = 0$ $x = 5, x = -3$ 7) $x^2 + 6x - 40 = 0$ $x = 4, x = -10$ 8) $x^2 - 9x - 36 = 0$ $x = 12, x = -3$

Name:	Date: ....

Topic	**Graphing Quadratic Functions**
Notes	✓ Quadratic functions in vertex form: $y = a(x-h)^2 + k$ where (h,k) is the vertex of the function. The axis of symmetry is $x = h$ ✓ Quadratic functions in standard form: $y = ax^2 + bx + c$ where $x = -\frac{b}{2a}$ is the value of x in the vertex of the function. ✓ To graph a quadratic function, first find the vertex, then substitute some values for x and solve for y.
Example	**Sketch the graph of $y = (x-2)^2 - 5$** **Solution:** The vertex of $y = (x-2)^2 - 5$ is $(2,5)$. Substitute zero for x and solve for y. $$y = (0-2)^2 - 5 = -1$$ The y-Intercept is $(0, -1)$ Now, you can simply graph the quadratic function. 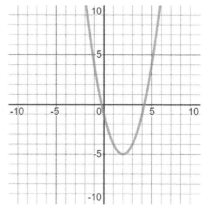
Your Turn!	1) $y = (x-4)^2 - 2$ 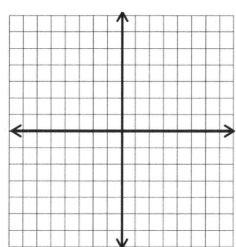 2) $y = 2(x+2)^2 - 3$ 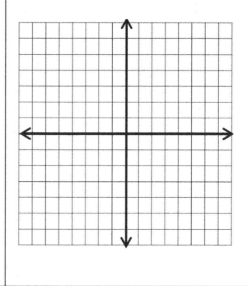

Name:	Date: ....

Topic	**Graphing Quadratic Functions- Answers**
Notes	✓ Quadratic functions in vertex form: $y = a(x-h)^2 + k$ where (h, k) is the vertex of the function. The axis of symmetry is $x = h$ ✓ Quadratic functions in standard form: $y = ax^2 + bx + c$ where $x = -\frac{b}{2a}$ is the value of x in the vertex of the function. ✓ To graph a quadratic function, first find the vertex, then substitute some values for x and solve for y.
Example	**Sketch the graph of $y = (x-2)^2 - 5$** **Solution:** The vertex of $y = (x-2)^2 - 5$ is $(2, 5)$. Substitute zero for x and solve for y. $$y = (0-2)^2 - 5 = -1$$ The y-Intercept is $(0, -1)$ Now, you can simply graph the quadratic function.
Your Turn!	1) $y = (x-4)^2 - 2$ 2) $y = 2(x+2)^2 - 3$

Name:	Date: ....

Topic	**Solving Quadratic Inequalities**
Notes	✓ A quadratic inequality is one that can be written in the standard form of $ax^2 + bx + c > 0$ (or substitute $<, \leq,$ or $\geq$ for $>$). ✓ Solving a quadratic inequality is like solving equations. We need to find the solutions (the zeroes). ✓ To solve quadratic inequalities, first find quadratic equations. Then choose a test value between zeroes. Finally, find interval(s), such as > 0 or < 0.
Example	**Solve quadratic inequality.** $x^2 + x - 12 > 0$ **Solution:** First solve $x^2 + x - 12 = 0$ by factoring. Then: $x^2 + x - 12 = 0 \rightarrow (x - 3)(x + 4) = 0$. The product of two expressions is 0. Then: $(x - 3) = 0 \rightarrow x = 3$ or $(x + 4) = 0 \rightarrow x = -4$. Now, choose a value between 3 and -4. Let's choose 0. Then: $x = 0 \rightarrow x^2 + x - 12 > 0 \rightarrow (0)^2 + (0) - 12 > 0 \rightarrow -12 > 0$ -12 is not greater than 0. Therefore, all values between 3 and -4 are NOT the solution of this quadratic inequality. The solution is: $x > 3$ and $x < -4$.
Your Turn!	1) $x^2 - 6x - 27 > 0$ _____ 3) $x^2 + x - 56 > 0$ _____ 5) $x^2 + 2x - 35 \leq 0$ _____	2) $x^2 + 13x + 42 < 0$ _____ 4) $x^2 - 15x + 54 < 0$ _____ 6) $x^2 - x - 72 \geq 0$ _____

Name:	Date: ....

Topic	**Solving Quadratic Inequalities - Answers**
Notes	✓ A quadratic inequality is one that can be written in the standard form of $ax^2 + bx + c > 0$ (or substitute $<, \leq$, or $\geq$ for $>$). ✓ Solving a quadratic inequality is like solving equations. We need to find the solutions (the zeroes). ✓ To solve quadratic inequalities, first find quadratic equations. Then choose a test value between zeroes. Finally, find interval(s), such as > 0 or < 0.
Example	**Solve quadratic inequality.** $x^2 + x - 12 > 0$ **Solution:** First solve $x^2 + x - 12 = 0$ by factoring. Then: $x^2 + x - 12 = 0 \rightarrow (x - 3)(x + 4) = 0$. The product of two expressions is 0. Then: $(x - 3) = 0 \rightarrow x = 3$ or $(x + 4) = 0 \rightarrow x = -4$. Now, choose a value between 3 and -4. Let's choose 0. Then: $x = 0 \rightarrow x^2 + x - 12 > 0 \rightarrow (0)^2 + (0) - 12 > 0 \rightarrow -12 > 0$ -12 is not greater than 0. Therefore, all values between 3 and -4 are NOT the solution of this quadratic inequality. The solution is: $x > 3$ and $x < -4$.

Your Turn!	1) $x^2 - 6x - 27 > 0$ $x < -3 \text{ or } x > 9$	2) $x^2 + 13x + 42 < 0$ $-7 < x < -6$
	3) $x^2 + x - 56 > 0$ $x < -8 \text{ or } x > 7$	4) $x^2 - 15x + 54 < 0$ $6 < x < 9$
	5) $x^2 + 2x - 35 \leq 0$ $-7 \leq x \leq 5$	6) $x^2 - x - 72 \geq 0$ $x \leq -8 \text{ or } x \geq 9$

Name:	....	Date:	....

Topic	**Graphing Quadratic Inequalities**
Notes	✓ A quadratic inequality is in the form $y > ax^2 + bx + c$ (or substitute $<$, $\leq$, or $\geq$ for $>$). ✓ To graph a quadratic inequality, start by graphing the quadratic parabola. Then fill in the region either inside or outside of it, depending on the inequality. ✓ Choose a testing point and check the solution section.
Example	**Sketch the graph of $y > x^2$** **Solution:** First, graph $y = x^2$ Since, the inequality sing is $>$, we need to use dash lines. Now, choose a testing point inside the parabola. Let's choose $(0,2)$. $y > x^2 \rightarrow 2 > (0)^2 \rightarrow 2 > 0$ This is true. So, inside the parabola is the solution section. 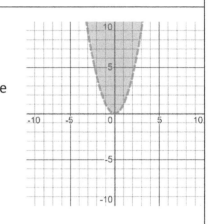
Your Turn!	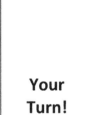 1) $y \leq x^2 + 4x + 5$ 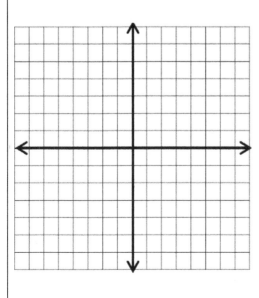 2) $y \leq x^2 + 2x - 3$ 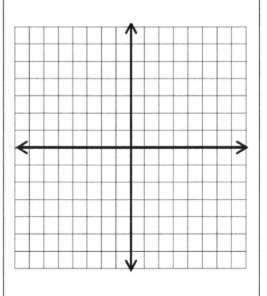

| Name: .... | Date: .... |

Topic	**Graphing Quadratic Inequalities- Answers**
Notes	✓ A quadratic inequality is in the form $y > ax^2 + bx + c$ (or substitute $<$, $\leq$, or $\geq$ for $>$). ✓ To graph a quadratic inequality, start by graphing the quadratic parabola. Then fill in the region either inside or outside of it, depending on the inequality. ✓ Choose a testing point and check the solution section.
Example	**Sketch the graph of $y > x^2$** **Solution:** First, graph $y = x^2$ Since, the inequality sing is $>$, we need to use dash lines. Now, choose a testing point inside the parabola. Let's choose $(0,2)$. $y > x^2 \rightarrow 2 > (0)^2 \rightarrow 2 > 0$ This is true. So, inside the parabola is the solution section.
Your Turn!	1) $y \leq x^2 + 4x + 5$ 2) $y \leq x^2 + 2x - 3$

Name:	Date: ....

Topic	**Adding and Subtracting Complex Numbers**
Notes	✓ A complex number is expressed in the form $a + bi$, where a and b are real numbers, and i, which is called an imaginary number, is a solution of the equation $x^2 = -1$ ✓ For adding complex numbers: $(a + bi) + (c + di) = (a + c) + (b + d)i$ ✓ For subtracting complex numbers: ✓ $(a + bi) - (c + di) = (a - c) + (b - d)i$
Example	**Solve:** $(14 + 7i) + (-5 - 3i)$ **Solution:** Remove parentheses: $(14 + 7i) + (-5 - 3i) \rightarrow 14 + 7i - 5 - 3i$ Combine like terms: $(14 - 5) + (7i - 3i) = 9 + 4i$

Your Turn!	1) $(5 - 3i) - (4 + i) =$ _____	2) $(2 + 6i) - (4 - 2i) =$ _____
	3) $(7 + 4i) - (5 - 6i) =$ _____	4) $(1 + 2i) + (5 - 7i) =$ _____
	5) $(-8 + 2i) - (5 - 3i) =$ _____	6) $(7 - 9i) - (3 + 5i) =$ _____
	7) $(-9 - 3i) - (9 - 10i) =$ _____	8) $(-12 - 4i) - (5 + 7i) =$ _____

| Name: .... | Date: .... |

Topic	Adding and Subtracting Complex Numbers - Answer
Notes	✓ A complex number is expressed in the form $a + bi$, where a and b are real numbers, and i, which is called an imaginary number, is a solution of the equation $x^2 = -1$ ✓ For adding complex numbers: $(a + bi) + (c + di) = (a + c) + (b + d)i$ ✓ For subtracting complex numbers: ✓ $(a + bi) - (c + di) = (a - c) + (b - d)i$
Example	**Solve:** $(14 + 7i) + (-5 - 3i)$ **Solution:** Remove parentheses: $(14 + 7i) + (-5 - 3i) \rightarrow 14 + 7i - 5 - 3i$ Combine like terms: $(14 - 5) + (7i - 3i) = 9 + 4i$

Your Turn!

1) $(5 - 3i) - (4 + i) =$ $1 - 4i$	2) $(2 + 6i) - (4 - 2i) =$ $-2 + 8i$
3) $(7 + 4i) - (5 - 6i) =$ $2 + 10i$	4) $(1 + 2i) + (5 - 7i) =$ $6 - 5i$
5) $(-8 + 2i) - (5 - 3i) =$ $-13 + 5i$	6) $(7 - 9i) - (3 + 5i) =$ $4 - 14i$
7) $(-9 - 3i) - (9 - 10i) =$ $-18 + 7i$	8) $(-12 - 4i) - (5 + 7i) =$ $-17 - 11i$

Name:	Date: ....

Topic	**Multiplying and Dividing Complex Numbers**	
Notes	✓ Multiplying complex numbers: $(a + bi) + (c + di) = (ac - bd) + (ad + bc)i$ ✓ Dividing complex numbers: $\frac{a+bi}{c+di} = \frac{a+bi}{c+di} \times \frac{c-di}{c-di} = \frac{ac+bd}{c^2+d^2} + \frac{bc-ad}{c^2+d^2}i$ ✓ Imaginary number rule: $i^2 = -1$	
Example	**Solve:** $\frac{8-2i}{2+i}$ **Solution:** Use the rule for dividing complex numbers: $$\frac{a+bi}{c+di} = \frac{a+bi}{c+di} \times \frac{c-di}{c-di} = \frac{ac+bd}{c^2+d^2} + \frac{bc-ad}{c^2+d^2}i$$ $$\frac{8-2i}{2+i} \times \frac{2-i}{2-i} = \frac{8\times(2)+(-2)(1)}{2^2+(1)^2} + \frac{-2\times(2)-(8)(1)}{2^2+(1)^2}i = \frac{14-12i}{5} =$$ $$\frac{14}{5} - \frac{12}{5}i$$	
Your Turn!	1) $(2 - 2i)(4 - i) =$ _____ 3) $(3 - i)(2 - 4i) =$ _____ 5) $\frac{5-i}{6+i} =$ _____	2) $(3 - 2i)(2 - i) =$ _____ 4) $(6 + i)(2 - 2i) =$ _____ 6) $\frac{7+2i}{3-2i} =$ _____

Name:	Date: ....

Topic	Multiplying and Dividing Complex Numbers-Answers
Notes	✓ Multiplying complex numbers: $(a+bi)+(c+di)=(ac-bd)+(ad+bc)i$ ✓ Dividing complex numbers: $\frac{a+bi}{c+di}=\frac{a+bi}{c+di}\times\frac{c-di}{c-di}=\frac{ac+bd}{c^2+d^2}+\frac{bc-ad}{c^2+d^2}i$ ✓ Imaginary number rule: $i^2=-1$
Example	**Solve:** $\frac{8-2i}{2+i}$ **Solution:** Use the rule for dividing complex numbers: $\frac{a+bi}{c+di}=\frac{a+bi}{c+di}\times\frac{c-di}{c-di}=\frac{ac+bd}{c^2+d^2}+\frac{bc-ad}{c^2+d^2}i$ $\frac{8-2i}{2+i}\times\frac{2-i}{2-i}=\frac{8\times(2)+(-2)(1)}{2^2+(1)^2}+\frac{-2\times(2)-(8)(1)}{2^2+(1)^2}i=\frac{14-12i}{5}=$ $\frac{14}{5}-\frac{12}{5}i$

Your Turn!

1) $(2-2i)(4-i)=$ $6-10i$	2) $(3-2i)(2-i)=$ $4-7i$
3) $(3-i)(2-4i)=$ $2-14i$	4) $(6+i)(2-2i)=$ $14-10i$
5) $\frac{5-i}{6+i}=$ $\frac{29}{37}-\frac{11}{37}i$	6) $\frac{7+2i}{3-2i}=$ $\frac{17}{13}+\frac{20}{13}i$

| Name: .... | Date: .... |

Topic	**Rationalizing Imaginary Denominators**
Notes	✓ Step 1: Find the conjugate (it's the denominator with different sign between the two terms. ✓ Step 2: Multiply numerator and denominator by the conjugate. ✓ Step 3: Simplify if needed.
Example	**Solve:** $\dfrac{6i}{3-3i}$ **Solution:** First, divide both sides of the fraction by 3: $\dfrac{6i}{3-3i} = \dfrac{2i}{1-i}$ Multiply both numerator and denominator by the conjugate $\dfrac{1+i}{1+i}$: $\dfrac{2i(1+i)}{(1-i)(1+i)} =$ Apply complex arithmetic rule: $(a+bi)(a-bi) = a^2 + b^2 \rightarrow (1-i)(1+i) =$ $1^2 + (-1)^2 = 2$, then: $\dfrac{2i(1+i)}{(1-i)(1+i)} = \dfrac{2i+2i^2}{2} = \dfrac{2i+2i^2}{2} = \dfrac{2i}{2} + \dfrac{2(-1)}{2} = -1+i$
Your Turn!	1) $\dfrac{2-i}{3i} =$ _____ 2) $\dfrac{4i+1}{2+i} =$ _____ 3) $\dfrac{6-3i}{3-i} =$ _____ 4) $\dfrac{8-3i}{2-i} =$ _____ 5) $\dfrac{-8+2i}{6-3i} =$ _____ 6) $\dfrac{-9+4i}{2-3i} =$ _____

Name:	Date: ....

Topic	**Rationalizing Imaginary Denominators - Answers**	
Notes	✓ Step 1: Find the conjugate (it's the denominator with different sign between the two terms. ✓ Step 2: Multiply numerator and denominator by the conjugate. ✓ Step 3: Simplify if needed.	
Example	**Solve:** $\dfrac{6i}{3-3i}$ **Solution:** First, divide both sides of the fraction by 3: $\dfrac{6i}{3-3i} = \dfrac{2i}{1-i}$ Multiply both numerator and denominator by the conjugate $\dfrac{1+i}{1+i}$: $\dfrac{2i(1+i)}{(1-i)(1+i)} =$ Apply complex arithmetic rule: $(a+bi)(a-bi) = a^2+b^2 \rightarrow (1-i)(1+i) =$ $1^2+(-1)^2 = 2$, then: $\dfrac{2i(1+i)}{(1-i)(1+i)} = \dfrac{2i+2i^2}{2} = \dfrac{2i+2i^2}{2} = \dfrac{2i}{2} + \dfrac{2(-1)}{2} = -1+i$	
Your **Turn!**	1) $\dfrac{2-i}{3i} =$ $-\dfrac{1}{3} - \dfrac{2}{3}i$	2) $\dfrac{4i+1}{2+i} =$ $\dfrac{6}{5} + \dfrac{7}{5}i$
	3) $\dfrac{6-3i}{3-i} =$ $\dfrac{21}{10} - \dfrac{3}{10}i$	4) $\dfrac{8-3i}{2-i} =$ $\dfrac{19}{5} + \dfrac{2}{5}i$
	5) $\dfrac{-8+2i}{6-3i} =$ $-\dfrac{6}{5} - \dfrac{4}{15}i$	6) $\dfrac{-9+4i}{2-3i} =$ $-\dfrac{30}{13} - \dfrac{19}{13}i$

Name:	Date: ....

Topic	**Function Notation and Evaluation**
Notes	✓ Functions are mathematical operations that assign unique outputs to given inputs. ✓ Function notation is the way a function is written. It is meant to be a precise way of giving information about the function without a rather lengthy written explanation. ✓ The most popular function notation is $f(x)$ which is read "f of x". ✓ To evaluate a function, plug in the input (the given value or expression) for the function's variable (place holder, x).
Example	**Evaluate**: $h(n) = 2n^2 - 2$, **find $h(2)$.** **Solution:** Substitute n with 2: Then: $h(n) = 2n^2 - 2 \rightarrow h(2) = 2(2)^2 - 2 = 8 - 2 \rightarrow h(2) = 6$

Your Turn!	1) $f(x) = x - 2$, find $f(-1)$	2) $g(x) = 2x + 4$, find $g(3)$
	3) $g(n) = 2n - 8$, find $g(-1)$	4) $h(n) = n^2 - 1$, find $h(-2)$
	5) $f(x) = x^2 + 12$, find $f(5)$	6) $g(x) = 2x^2 - 9$, find $g(-2)$
	7) $w(x) = 2x^2 - 4x$, find $w(2n)$	8) $p(x) = 4x^3 - 10x$, find $p(-3a)$

| Name: .... | Date: .... |

Topic	**Function Notation and Evaluation - Answers**
Notes	✓ Functions are mathematical operations that assign unique outputs to given inputs. ✓ Function notation is the way a function is written. It is meant to be a precise way of giving information about the function without a rather lengthy written explanation. ✓ The most popular function notation is $f(x)$ which is read "f of x". ✓ To evaluate a function, plug in the input (the given value or expression) for the function's variable (place holder, x).
Example	**Evaluate**: $h(n) = 2n^2 - 2$, find $h(2)$. **Solution:** Substitute n with 2: Then: $h(n) = 2n^2 - 2 \rightarrow h(2) = 2(2)^2 - 2 = 8 - 2 \rightarrow h(2) = 6$

Your Turn!	
1) $f(x) = x - 2$, find $f(-1)$ $f(-1) = -3$	2) $g(x) = 2x + 4$, find $g(3)$ $g(3) = 10$
3) $g(n) = 2n - 8$, find $g(-1)$ $g(-1) = -10$	4) $h(n) = n^2 - 1$, find $h(-2)$ $h(-2) = 3$
5) $f(x) = x^2 + 12$, find $f(5)$ $f(5) = 37$	6) $g(x) = 2x^2 - 9$, find $g(-2)$ $g(-2) = -1$
7) $w(x) = 2x^2 - 4x$, find $w(2n)$ $w(2n) = 8n^2 - 8n$	8) $p(x) = 4x^3 - 10x$, find $p(-3a)$ $p(-3a) = -108a^3 + 30a$

Name:	Date: ....

Topic	Adding and Subtracting Functions
Notes	✓ Just like we can add and subtract numbers and expressions, we can add or subtract two functions and simplify or evaluate them. The result is a new function. ✓ For two functions $f(x)$ and $g(x)$, we can create two new functions: $(f + g)(x) = f(x) + g(x)$ and $(f - g)(x) = f(x) - g(x)$

Example	$g(a) = 2a - 5$, $f(a) = a + 8$, Find: $(g + f)(a)$
	
	    Solution: $(g + f)(a) = g(a) + f(a)$
	
	    Then: $(g + f)(a) = (2a - 5) + (a + 8) = 3a + 3$

Your Turn!	1) $g(x) = x - 2$
	
	    $h(x) = 2x + 6$
	    Find: $(h + g)(3)$
	
	    _____
		2) $f(x) = 3x + 2$
		
		    $g(x) = -x - 6$
		    Find: $(f + g)(2)$
		
		    _____

	3) $f(x) = 5x + 8$
	
	    $g(x) = 3x - 12$
	    Find: $(f - g)(-2)$
	
	    _____
		4) $h(x) = 2x^2 - 10$
		
		    $g(x) = 3x + 12$
		    Find: $(h + g)(3)$
		
		    _____

	5) $g(x) = 12x - 8$
	
	    $h(x) = 3x^2 + 14$
	    Find: $(h - g)(x)$
	
	    _____
		6) $h(x) = -2x^2 - 18$
		
		    $g(x) = 4x^2 + 15$
		    Find: $(h - g)(a)$
		
		    _____

Name:	Date: ....

Topic	**Adding and Subtracting Functions - Answers**
Notes	✓ Just like we can add and subtract numbers and expressions, we can add or subtract two functions and simplify or evaluate them. The result is a new function. ✓ For two functions $f(x)$ and $g(x)$, we can create two new functions: $(f + g)(x) = f(x) + g(x)$ and $(f - g)(x) = f(x) - g(x)$
Example	$g(a) = 2a - 5, f(a) = a + 8$, **Find:** $(g + f)(a)$ **Solution:** $(g + f)(a) = g(a) + f(a)$ Then: $(g + f)(a) = (2a - 5) + (a + 8) = 3a + 3$

Your Turn!		
	1) $g(x) = x - 2$ $h(x) = 2x + 6$ Find: $(h + g)(3)$ 13	2) $f(x) = 3x + 2$ $g(x) = -x - 6$ Find: $(f + g)(2)$ 0
	3) $f(x) = 5x + 8$ $g(x) = 3x - 12$ Find: $(f - g)(-2)$ 16	4) $h(x) = 2x^2 - 10$ $g(x) = 3x + 12$ Find: $(h + g)(3)$ 29
	5) $g(x) = 12x - 8$ $h(x) = 3x^2 + 14$ Find: $(h - g)(x)$ $3x^2 - 12x + 22$	6) $h(x) = -2x^2 - 18$ $g(x) = 4x^2 + 15$ Find: $(h - g)(a)$ $-6a^2 - 33$

Name:	Date: ....

Topic	**Multiplying and Dividing Functions**
Notes	✓ Just like we can multiply and divide numbers and expressions, we can multiply and divide two functions and simplify or evaluate them. ✓ For two functions $f(x)$ and $g(x)$, we can create two new functions: $(f \cdot g)(x) = f(x) \cdot g(x)$ and $\left(\frac{f}{g}\right)(x) = \frac{f(x)}{g(x)}$
Example	$g(x) = x + 5, f(x) = x - 3$, **Find:** $(g \cdot f)(2)$ **Solution:** $(g \cdot f)(x) = g(x) \cdot f(x) = (x + 5)(x - 3) = x^2 - 3x + 5x - 15 = x^2 + 2x - 15$ Substitute x with 2: $(g \cdot f)(x) = (2)^2 + 2(2) - 15 = 4 + 4 - 15 = -7$

Your Turn!	1) $g(x) = x - 5$ $h(x) = x + 6$ Find: $(g \cdot h)(-1)$ _____	2) $f(x) = 2x + 2$ $g(x) = -x - 6$ Find: $\left(\frac{f}{g}\right)(-2)$ _____
	3) $f(x) = 5x + 3$ $g(x) = 2x - 4$ Find: $\left(\frac{f}{g}\right)(5)$ _____	4) $h(x) = x^2 - 2$ $g(x) = x + 4$ Find: $(g \cdot h)(3)$ _____
	5) $g(x) = 4x - 12$ $h(x) = x^2 + 4$ Find: $(g \cdot h)(-2)$ _____	6) $f(x) = 3x^2 - 8$ $g(x) = 4x + 6$ Find: $\left(\frac{f}{g}\right)(-4)$ _____

Name:	Date: ....

Topic	**Multiplying and Dividing Functions - Answers**
Notes	✓ Just like we can multiply and divide numbers and expressions, we can multiply and divide two functions and simplify or evaluate them. ✓ For two functions $f(x)$ and $g(x)$, we can create two new functions: $(f \cdot g)(x) = f(x) \cdot g(x)$ and $\left(\dfrac{f}{g}\right)(x) = \dfrac{f(x)}{g(x)}$
Example	$g(x) = x + 5, f(x) = x - 3$, Find: $(g \cdot f)(2)$ **Solution:** $(g \cdot f)(x) = g(x) \cdot f(x) = (x + 5)(x - 3) = x^2 - 3x + 5x - 15 = x^2 + 2x - 15$ Substitute x with 2: $(g \cdot f)(x) = (2)^2 + 2(2) - 15 = 4 + 4 - 15 = -7$

Your Turn!		
	1) $g(x) = x - 5$ $h(x) = x + 6$ Find: $(g \cdot h)(-1)$ $(g \cdot h)(-1) = -30$	2) $f(x) = 2x + 2$ $g(x) = -x - 6$ Find: $\left(\dfrac{f}{g}\right)(-2)$ $\left(\dfrac{f}{g}\right)(-2) = \dfrac{1}{2}$
	3) $f(x) = 5x + 3$ $g(x) = 2x - 4$ Find: $\left(\dfrac{f}{g}\right)(5)$ $\left(\dfrac{f}{g}\right)(5) = \dfrac{14}{3}$	4) $h(x) = x^2 - 2$ $g(x) = x + 4$ Find: $(g \cdot h)(3)$ $(g \cdot h)(3) = 49$
	5) $g(x) = 4x - 12$ $h(x) = x^2 + 4$ Find: $(g \cdot h)(-2)$ $(g \cdot h)(-2) = -160$	6) $f(x) = 3x^2 - 8$ $g(x) = 4x + 6$ Find: $\left(\dfrac{f}{g}\right)(-4)$ $\left(\dfrac{f}{g}\right)(-4) = -4$

| Name: .... | Date: .... |

Topic	Composition of Functions
Notes	✓ "Composition of functions" simply means combining two or more functions in a way where the output from one function becomes the input for the next function. ✓ The notation used for composition is: $(fog)(x) = f(g(x))$ and is read "f composed with g of x" or "f of g of x".
Example	**Using $f(x) = x - 8$ and $g(x) = x + 2$, find:** $(fog)(3)$ **Solution:** $(f \ o \ g)(x) = f(g(x))$ Then: $(fog)(x) = f(g(x)) = f(x + 2) = x + 2 - 8 = x - 6$ Substitute x with 3: $(fog)(3) = f(g(3)) = 3 - 6 = -3$

Your Turn!	1) $f(x) = 2x$ $g(x) = x + 3$ Find: $(fog)(2)$ _____	2) $f(x) = x + 2$ $g(x) = x - 6$ Find: $(fog)(-1)$ _____
	3) $f(x) = 3x$ $g(x) = x + 4$ Find: $(gof)(4)$ _____	4) $h(x) = 2x - 2$ $g(x) = x + 4$ Find: $(goh)(2)$ _____
	5) $f(x) = 2x - 8$ $g(x) = x + 10$ Find: $(fog)(-2)$ _____	6) $f(x) = x^2 - 8$ $g(x) = 2x + 3$ Find: $(gof)(4)$ _____

Name: ....

Date: ....

Topic	Composition of Functions - Answers
Notes	✓ "Composition of functions" simply means combining two or more functions in a way where the output from one function becomes the input for the next function. ✓ The notation used for composition is: $(fog)(x) = f(g(x))$ and is read "f composed with g of x" or "f of g of x".
Example	**Using $f(x) = x - 8$ and $g(x) = x + 2$, find: $(fog)(3)$** **Solution:** $(fog)(x) = f(g(x))$ Then: $(fog)(x) = f(g(x)) = f(x+2) = x + 2 - 8 = x - 6$ Substitute x with 3: $(fog)(3) = f(g(3)) = 3 - 6 = -3$
Your Turn!	1) $f(x) = 2x$ $g(x) = x + 3$ Find: $(fog)(2)$ 10 <hr> 3) $f(x) = 3x$ $g(x) = x + 4$ Find: $(gof)(4)$ 16 <hr> 5) $f(x) = 2x - 8$ $g(x) = x + 10$ Find: $(fog)(-2)$ 8

Columns 2 continued (right side of Your Turn):

2) $f(x) = x + 2$

$g(x) = x - 6$
Find: $(fog)(-1)$

-5

4) $h(x) = 2x - 2$

$g(x) = x + 4$
Find: $(goh)(2)$

6

6) $f(x) = x^2 - 8$

$g(x) = 2x + 3$
Find: $(gof)(4)$

19

| Name: .... | Date: .... |

Topic	Function Inverses

Notes	✓ An inverse function is a function that reverses another function: if the function f applied to an input x gives a result of y, then applying its inverse function g to y gives the result x. $f(x) = y$ if and only if $g(y) = x$

Examples	**1) Find the inverse of $f(x) = 4x + 2$** Solution: First, replace $f(x)$ with y: $y = 4x + 2$ Next, replace all x's with y and all y's with x: $x = 4y + 2$ Now, solve for y: $x = 4y + 2 \rightarrow x - 2 = 4y \rightarrow \frac{1}{4}x - \frac{1}{2} = y$ Finally replace y with $f^{-1}(x)$: $f^{-1}(x) = \frac{1}{4}x - \frac{1}{2}$ **2) Find the inverse of $h(x) = \frac{x+1}{2}$** Solution: $h(x) = \frac{x+1}{2} \rightarrow y = \frac{x+1}{2}$, replace all x's with y and all y's with x: $x = \frac{y+1}{2} \rightarrow 2x = y + 1 \rightarrow 2x - 1 = y \rightarrow h^{-1}(x) = 2x - 1$

	Find the inverse of each function.

Your Turn!	1) $f(x) = -\frac{1}{x} - 9$
	     $f^{-1}(x) = $ _____
		2) $g(x) = \sqrt{x} - 2$
		     $g^{-1}(x) = $ _____

	3) $h(x) = -\frac{5}{x+3}$
	     $h^{-1}(x) = $ _____
		4) $f(x) = 6x + 6$
		     $f^{-1}(x) = $ _____

| **Name:** .... | **Date:** .... |

Topic	**Function Inverses - Answers**
Notes	✓ An inverse function is a function that reverses another function: if the function f applied to an input x gives a result of y, then applying its inverse function g to y gives the result x. $f(x) = y$ if and only if $g(y) = x$
Examples	**1) Find the inverse of $f(x) = 4x + 2$** Solution: First, replace $f(x)$ with y: $y = 4x + 2$ Next, replace all x's with y and all y's with x: $x = 4y + 2$ Now, solve for y: $x = 4y + 2 \rightarrow x - 2 = 4y \rightarrow \frac{1}{4}x - \frac{1}{2} = y$ Finally replace y with $f^{-1}(x)$: $f^{-1}(x) = \frac{1}{4}x - \frac{1}{2}$ **2) Find the inverse of $h(x) = \frac{x+1}{2}$** Solution: $h(x) = \frac{x+1}{2} \rightarrow y = \frac{x+1}{2}$, replace all x's with y and all y's with x: $x = \frac{y+1}{2} \rightarrow 2x = y + 1 \rightarrow 2x - 1 = y \rightarrow h^{-1}(x) = 2x - 1$
Your Turn!	**Find the inverse of each function.**

1) $f(x) = -\dfrac{1}{x} - 9$ $f^{-1}(x) = -\dfrac{1}{x+9}$	2) $g(x) = \sqrt{x} - 2$ $g^{-1}(x) = x^2 + 4x + 4$
3) $h(x) = -\dfrac{5}{x+3}$ $h^{-1}(x) = -\dfrac{5}{x} - 3$	4) $f(x) = 6x + 6$ $f^{-1}(x) = \dfrac{x-6}{6}$

Name:	Date: ....

Topic	**Simplifying Radical Expressions**
Notes	✓ Find the prime factors of the numbers or expressions inside the radical. ✓ Use radical properties to simplify the radical expression: $$\sqrt[n]{x^a} = x^{\frac{a}{n}}, \sqrt[n]{xy} = x^{\frac{1}{n}} \times y^{\frac{1}{n}}, \sqrt[n]{\frac{x}{y}} = \frac{x^{\frac{1}{n}}}{y^{\frac{1}{n}}}, \text{ and } \sqrt[n]{x} \times \sqrt[n]{y} = \sqrt[n]{xy}$$
Example	**Evaluate.** $\sqrt{64} \times \sqrt{y^2} =$ First factor the numbers: $64 = 8^2$ Then: $\sqrt{64} \times \sqrt{y^2} = \sqrt{8^2} \times \sqrt{y^2}$ Now use radical rule: $\sqrt[n]{a^n} = a$, Then: $\sqrt{8^2} \times \sqrt{y^2} = 8 \times y = 8y$
Your Turn!	1) Evaluate. $\sqrt{49} =$ _____ 2) Evaluate. $\sqrt{4} \times \sqrt{81} =$ _____ 3) Evaluate. $\sqrt{16} \times \sqrt{4x^2} =$ _____ 4) Evaluate. $\sqrt{289} =$ _____ 5) Evaluate. $\sqrt{25b^4} =$ _____ 6) Evaluate. $\sqrt{9} \times \sqrt{x^2} =$ _____

| Name: .... | Date: .... |

Topic	**Simplifying Radical Expressions - Answers**
Notes	✓ Find the prime factors of the numbers or expressions inside the radical. ✓ Use radical properties to simplify the radical expression: $$\sqrt[n]{x^a} = x^{\frac{a}{n}}, \sqrt[n]{xy} = x^{\frac{1}{n}} \times y^{\frac{1}{n}}, \sqrt[n]{\frac{x}{y}} = \frac{x^{\frac{1}{n}}}{y^{\frac{1}{n}}}, \text{ and } \sqrt[n]{x} \times \sqrt[n]{y} = \sqrt[n]{xy}$$
Example	**Evaluate.** $\sqrt{64} \times \sqrt{y^2} =$ First factor the numbers: $64 = 8^2$ Then: $\sqrt{64} \times \sqrt{y^2} = \sqrt{8^2} \times \sqrt{y^2}$ Now use radical rule: $\sqrt[n]{a^n} = a$, Then: $\sqrt{8^2} \times \sqrt{y^2} = 8 \times y = 8y$
Your Turn!	1) Evaluate. $\sqrt{49} = 7$ 2) Evaluate. $\sqrt{4} \times \sqrt{81} = 18$ 3) Evaluate. $\sqrt{16} \times \sqrt{4x^2} = 8x$ 4) Evaluate. $\sqrt{289} = 17$ 5) Evaluate. $\sqrt{25b^4} = 5b^2$ 6) Evaluate. $\sqrt{9} \times \sqrt{x^2} = 3x$

Name:	Date: ....

Topic	**Adding and Subtracting Radical Expressions**	
Notes	✓ Only numbers that have the same radical part can be added or subtracted. ✓ Remember, combining "unlike" radical terms is not possible. ✓ For number with the same radical part, just add or subtract factors outside the radicals.	
Examples	1) **Simplify.** $6\sqrt{5} + 3\sqrt{5}$ Add like terms: $6\sqrt{5} + 3\sqrt{5} = 9\sqrt{5}$ 2) **Simplify.** $5\sqrt{7} - 3\sqrt{7}$ Combine like terms: $5\sqrt{7} - 3\sqrt{7} = 2\sqrt{7}$	
Your Turn!	1) Simplify: $\sqrt{6} + 6\sqrt{6} =$ _____	2) Simplify: $9\sqrt{8} - 6\sqrt{2} =$ _____
	3) Simplify: $-\sqrt{7} - 5\sqrt{7} =$ _____	4) Simplify: $10\sqrt{2} + 3\sqrt{18} =$ _____
	5) Simplify: $\sqrt{12} - 6\sqrt{3} =$ _____	6) Simplify: $-2\sqrt{x} + 6\sqrt{x} =$ _____

Name:	Date: ....

Topic	Adding and Subtracting Radical Expressions - Answers
Notes	✓ Only numbers that have the same radical part can be added or subtracted. ✓ Remember, combining "unlike" radical terms is not possible. ✓ For number with the same radical part, just add or subtract factors outside the radicals.
Examples	1) **Simplify.** $6\sqrt{5} + 3\sqrt{5}$ Add like terms: $6\sqrt{5} + 3\sqrt{5} = 9\sqrt{5}$ 2) **Simplify.** $5\sqrt{7} - 3\sqrt{7}$ Combine like terms: $5\sqrt{7} - 3\sqrt{7} = 2\sqrt{7}$

Your Turn!		
1) Simplify: $\sqrt{6} + 6\sqrt{6} = 7\sqrt{6}$	2) Simplify: $9\sqrt{8} - 6\sqrt{2} = 12\sqrt{2}$
3) Simplify: $-\sqrt{7} - 5\sqrt{7} = -6\sqrt{7}$	4) Simplify: $10\sqrt{2} + 3\sqrt{18} = 19\sqrt{2}$
5) Simplify: $\sqrt{12} - 6\sqrt{3} = -4\sqrt{3}$	6) Simplify: $-2\sqrt{x} + 6\sqrt{x} = 4\sqrt{x}$

Name:	Date: ....

Topic	**Multiplying Rational Expressions**
Notes	✓ Multiplying rational expressions is the same as multiplying fractions. First, multiply numerators and then multiply denominators. Then, simplify as needed.
Examples	**1) Solve:** $\frac{x+5}{x-1} \times \frac{x-1}{3} =$ Multiply fractions: $\frac{x+5}{x-1} \times \frac{x-1}{3} = \frac{(x+5)(x-1)}{3(x-1)}$ Cancel the common factor: $(x-1)$, then: $\frac{(x+5)(x-1)}{3(x-1)} = \frac{(x+5)}{3}$ **2) Solve:** $\frac{x-5}{x+4} \times \frac{2x+8}{x-5} =$ Multiply fractions: $\frac{x-5}{x+4} \times \frac{2x+8}{x-5} = \frac{(x-5)(2x+8)}{(x+4)(x-5)}$ Cancel the common factor: $\frac{(x-5)(2x+8)}{(x+4)(x-5)} = \frac{(2x+8)}{(x+4)}$ Factor $2x+8 = 2(x+4)$, Then: $\frac{2(x+4)}{(x+4)} = 2$
Your Turn!	1) $\frac{20x^3}{3} \times \frac{15}{4x} =$ _____ 2) $\frac{x+6}{4} \times \frac{16}{x+6} =$ _____ 3) $\frac{x+10}{4x} \times \frac{3x}{7x+70} =$ _____ 4) $\frac{x+8}{x+6} \times \frac{x-6}{4x+32} =$ _____

Name:	Date: ....

Topic	**Multiplying Rational Expressions - Answers**
Notes	✓ Multiplying rational expressions is the same as multiplying fractions. First, multiply numerators and then multiply denominators. Then, simplify as needed.
Examples	**1)** Solve: $\frac{x+5}{x-1} \times \frac{x-1}{3} =$ Multiply fractions: $\frac{x+5}{x-1} \times \frac{x-1}{3} = \frac{(x+5)(x-1)}{3(x-1)}$ Cancel the common factor: $(x-1)$, then: $\frac{(x+5)(x-1)}{3(x-1)} = \frac{(x+5)}{3}$ **2)** Solve: $\frac{x-5}{x+4} \times \frac{2x+8}{x-5} =$ Multiply fractions: $\frac{x-5}{x+4} \times \frac{2x+8}{x-5} = \frac{(x-5)(2x+8)}{(x+4)(x-5)}$ Cancel the common factor: $\frac{(x-5)(2x+8)}{(x+4)(x-5)} = \frac{(2x+8)}{(x+4)}$ Factor $2x+8 = 2(x+4)$, Then: $\frac{2(x+4)}{(x+4)} = 2$

Your Turn!	**1)** $\frac{20x^3}{3} \times \frac{15}{4x} = 25x^2$	**2)** $\frac{x+6}{4} \times \frac{16}{x+6} = 4$
	3) $\frac{x+10}{4x} \times \frac{3x}{7x+70} = \frac{3}{28}$	**4)** $\frac{x+8}{x+6} \times \frac{x-6}{4x+32} = \frac{x-6}{4(x+6)}$

Name:	Date: ....

Topic	**Simplifying Radical Expressions Involving Fractions**
Notes	✓ Radical expressions cannot be in the denominator. (number in the bottom) ✓ To get rid of the radical in the denominator, multiply both numerator and denominator by the radical in the denominator. ✓ If there is a radical and another integer in the denominator, multiply both numerator and denominator by the conjugate of the denominator. ✓ The conjugate of $a + b$ is $a - b$ and vice versa.
Example	**Simplify** $\dfrac{1}{\sqrt{5} - 2}$ Multiply by the conjugate: $\dfrac{\sqrt{5}+2}{\sqrt{5}+2} \rightarrow \dfrac{1}{\sqrt{5}-2} \times \dfrac{\sqrt{5}+2}{\sqrt{5}+2}$ $\left(\sqrt{5} - 2\right)\left(\sqrt{5} + 2\right) = 1$ then: $\dfrac{1}{\sqrt{5}-2} \times \dfrac{\sqrt{5}+2}{\sqrt{5}+2} = \dfrac{\left(\sqrt{5}+2\right)}{1} =$ $\sqrt{5} + 2$

Your Turn!	1) Simplify: $\dfrac{1+\sqrt{5}}{1-\sqrt{3}} = $ _____	2) Simplify: $\dfrac{2+\sqrt{6}}{\sqrt{2}-\sqrt{5}} = $ _____
	3) Simplify: $\dfrac{\sqrt{7}}{\sqrt{6}-\sqrt{3}} = $ _____	4) Simplify: $\dfrac{\sqrt{8a}}{a^5} = $ _____

Name:	Date: ....

Topic	**Simplifying Radical Expressions Involving Fractions - Answers**
Notes	✓ Radical expressions cannot be in the denominator. (number in the bottom) ✓ To get rid of the radical in the denominator, multiply both numerator and denominator by the radical in the denominator. ✓ If there is a radical and another integer in the denominator, multiply both numerator and denominator by the conjugate of the denominator. ✓ The conjugate of $a + b$ is $a - b$ and vice versa.
Example	**Simplify** $\dfrac{1}{\sqrt{5}- 2}$ Multiply by the conjugate: $\dfrac{\sqrt{5}+2}{\sqrt{5}+2} \rightarrow \dfrac{1}{\sqrt{5}-2} \times \dfrac{\sqrt{5}+2}{\sqrt{5}+2}$ $\left(\sqrt{5} - 2\right)\left(\sqrt{5} + 2\right) = 1$ then: $\dfrac{1}{\sqrt{5}-2} \times \dfrac{\sqrt{5}+2}{\sqrt{5}+2} = \dfrac{\left(\sqrt{5}+2\right)}{1} =$ $\sqrt{5} + 2$

Your Turn!	1) Simplify: $\dfrac{1+\sqrt{5}}{1-\sqrt{3}} =$ $\dfrac{\left(1+\sqrt{5}\right)\left(1+\sqrt{3}\right)}{2}$	2) Simplify: $\dfrac{2+\sqrt{6}}{\sqrt{2}-\sqrt{5}} =$ $\dfrac{2\sqrt{2}+2\sqrt{5}+2\sqrt{3}+\sqrt{30}}{3}$
	3) Simplify: $\dfrac{\sqrt{7}}{\sqrt{6}-\sqrt{3}} = \dfrac{\sqrt{7}\left(\sqrt{6}+\sqrt{3}\right)}{3}$	4) Simplify: $\dfrac{\sqrt{8a}}{a^5} = \dfrac{2\sqrt{2}}{a^4\sqrt{a}}$

| Name: .... | Date: .... |

Topic	Radical Equations
Notes	✓ Isolate the radical on one side of the equation. ✓ Square both sides of the equation to remove the radical ✓ Solve the equation for the variable ✓ Plugin the answer into the original equation to avoid extraneous values.
Example	**Solve $\sqrt{x} - 8 = -3$** Add 8 to both sides: $\sqrt{x} = 5$ Square both sides: $\left(\sqrt{x}\right)^2 = 5^2 \rightarrow x = 25$ Substitute x by 25 in the original equation and check the answer: $$x = 25 \rightarrow \sqrt{x} - 8 = \sqrt{25} - 8 = -3$$ So, the value of 25 for x is correct.

Your Turn!	1) Solve: $2\sqrt{2x - 4} = 8$ _____	2) Solve: $9 = \sqrt{4x - 1}$ _____
	3) Solve: $\sqrt{x} + 6 = 11$ _____	4) Solve: $\sqrt{5x} = \sqrt{x + 3}$ _____

Name: ....

Date: ....

Topic	Radical Equations - Answers
Notes	✓ Isolate the radical on one side of the equation. ✓ Square both sides of the equation to remove the radical ✓ Solve the equation for the variable ✓ Plugin the answer into the original equation to avoid extraneous values.
Example	**Solve $\sqrt{x} - 8 = -3$** Add 8 to both sides: $\sqrt{x} = 5$ Square both sides: $\left(\sqrt{x}\right)^2 = 5^2 \rightarrow x = 25$ Substitute x by 25 in the original equation and check the answer: $$x = 25 \rightarrow \sqrt{x} - 8 = \sqrt{25} - 8 = -3$$ So, the value of 25 for x is correct.

Your Turn!	1) Solve: $2\sqrt{2x - 4} = 8$ $x = 10$	2) Solve: $9 = \sqrt{4x - 1}$ $x = 20.5$
	3) Solve: $\sqrt{x} + 6 = 11$ $x = 25$	4) Solve: $\sqrt{5x} = \sqrt{x + 3}$ $x = \dfrac{3}{4}$

Name:	Date: ....

Topic	**Domain and Range of Radical Functions**
Notes	✓ To find the domain of the function, find all possible values of the variable inside radical. ✓ Remember that having a negative number under the square root symbol is not possible. (For cubic roots, we can have negative numbers) ✓ To find the range, plugin the minimum and maximum values of the variable inside radical.
Example	**Find the domain and range of the radical function.** $$y = \sqrt{x - 8} + 5$$ For domain: Find non-negative values for radicals: $x - 8 \geq 0$ Then solve for x: $x - 8 \geq 0 \rightarrow x \geq 8$ Domain: $x \geq 8$ For range: the range of a radical function of the form $c\sqrt{ax + b} + k$ is $f(x) \geq k$ $k = 5$, Then: $f(x) \geq 5$

Your Turn!	1) Identify the Domain and Range: $$y = \sqrt{x + 1}$$	2) Identify the Domain and Range: $$y = \sqrt{x - 2} + 6$$
	3) Sketch the graph of function: $$y = 2\sqrt{x} + 1$$ 	4) Sketch the graph of function: $$y = \sqrt{x} + 5$$

Name: ....

Date: ....

Topic	Domain and Range of Radical Functions - Answers
Notes	✓ To find the domain of the function, find all possible values of the variable inside radical. ✓ Remember that having a negative number under the square root symbol is not possible. (For cubic roots, we can have negative numbers) ✓ To find the range, plugin the minimum and maximum values of the variable inside radical.
Example	**Find the domain and range of the radical function.** $$y = \sqrt{x - 8} + 5$$ For domain: Find non-negative values for radicals: $x - 8 \geq 0$ Then solve for x: $x - 8 \geq 0 \rightarrow x \geq 8$ Domain: $x \geq 8$ For range: the range of a radical function of the form $c\sqrt{ax + b} + k$ is $f(x) \geq k$ $k = 5$, Then: $f(x) \geq 5$

Your Turn!	1) Identify the Domain and Range: $y = \sqrt{x + 1}, x \geq -1, y \geq 0$	2) Identify the Domain and Range: $y = \sqrt{x - 2} + 6, x \geq 2, y \geq 6$
	3) Sketch the graph of function: $y = 2\sqrt{x} + 1$ 	4) Sketch the graph of function: $y = \sqrt{x} + 5$

Name:	Date: ....

Topic	Properties of Logarithms
Notes	✓ Learn some logarithms properties: $a^{\log_a b} = b$ $\qquad$ $\log_a(x \cdot y) = \log_a x + \log_a y$ $\log_a 1 = 0$ $\qquad$ $\log_a \frac{x}{y} = \log_a x - \log_a y$ $\log_a a = 1$ $\qquad$ $\log_{a^k} x = \frac{1}{k} \log_a x \,, for \; k \neq 0$ $\log_a \frac{1}{x} = -\log_a x$ $\qquad$ $\log_a x^p = p \log_a x$ $\log_a x = \frac{1}{\log_x a}$ $\qquad$ $\log_a x = \log_{a^c} x^c$
Example	Condense this expression to a single logarithm. $$\log_b 2 - \log_b 7$$ **Solution:** Use log rule: $\log_a x - \log_a y = \log_a \frac{x}{y}$ Then: $\log_b 2 - \log_b 7 = \log_b \frac{2}{7}$
Your Turn!	Condense this expression to a single logarithm. 1) $\log_a 5 - \log_a 8 =$ _____ $\qquad$ 2) $\log_x 3 - \log_x 5 =$ _____ 3) $\log_b 2 + \log_b 3 =$ _____ $\qquad$ 4) $\log_a 7 + \log_a 2 =$ _____

Name:	**Date:** ....

Topic	**Properties of Logarithms - Answers**
Notes	✓ Learn some logarithms properties: $a^{\log_a b} = b$ $\qquad\qquad\qquad$ $\log_a(x \cdot y) = \log_a x + \log_a y$ $\log_a 1 = 0$ $\qquad\qquad\qquad$ $\log_a \frac{x}{y} = \log_a x - \log_a y$ $\log_a a = 1$ $\qquad\qquad\qquad$ $\log_{a^k} x = \frac{1}{k} \log_a x$, $for\ k \neq 0$ $\log_a \frac{1}{x} = -\log_a x$ $\qquad\qquad$ $\log_a x^p = p \log_a x$ $\log_a x = \frac{1}{\log_x a}$ $\qquad\qquad\quad$ $\log_a x = \log_{a^c} x^c$
Example	**Condense this expression to a single logarithm.** $$\log_b 2 - \log_b 7$$ **Solution:** Use log rule: $\log_a x - \log_a y = \log_a \frac{x}{y}$ Then: $\log_b 2 - \log_b 7 = \log_b \frac{2}{7}$
Your Turn!	**Condense this expression to a single logarithm.** 1) $\log_a 5 - \log_a 8 =$ $\qquad$ 2) $\log_x 3 - \log_x 5 =$ $\qquad\qquad \log_a \frac{5}{8}$ $\qquad\qquad\qquad\qquad \log_x \frac{3}{5}$ 3) $\log_b 2 + \log_b 3 =$ $\qquad$ 4) $\log_a 7 + \log_a 2 =$ $\qquad\qquad \log_b (6)$ $\qquad\qquad\qquad\qquad \log_a (14)$

Name:	Date: ....

Topic	**Evaluating Logarithm**

Notes	✓ Logarithm is another way of writing exponent. $log_b{}^y = x$ is equivalent to $y = b^x$.
	
	✓ Learn some logarithms rules: ($a > 0, a \neq 0, M > 0, N > 0$, and k is a real number.)
	
	    Rule 1: $log_a(M.N) = log_aM + log_aN$
	
	    Rule 2: $log_a\frac{M}{N} = log_aM - log_aN$
	
	    Rule 3: $log_a(M)^k = klog_aM$
	
	    Rule 4: $log_aa = 1$
	
	    Rule 5: $log_a 1 = 0$
	
	    Rule 6: $a^{log_ak} = k$

Example	**Evaluate $3log_2(8)$**
	
	    Solution: $8 = 2^3$, then $log_2(8) = log_2(2)^3$
	
	    Use log rule: $log_a(M)^k = klog_a(M) \rightarrow log_2(2)^3 = 3log_2(2)$
	
	    Use log rule: $log_a(a) = 1 \rightarrow 3 \times 3log_2{}^2 = 3 \times 3 = 9$

Your Turn!	1) $2log_3(27) = $ _____ 2) $3log_4(256) = $ _____ 3) $\frac{1}{2}log_3(9) = $ _____ 4) $3log_5(25) = $ _____ 5) $6log_3(3) = $ _____ 6) $4log_6(1) = $ _____

Name:	Date: ....

Topic	**Evaluating Logarithm - Answers**
Notes	✓ Logarithm is another way of writing exponent. $log_b{}^y = x$ is equivalent to $y = b^x$. ✓ Learn some logarithms rules: ($a > 0, a \neq 0, M > 0, N > 0$, and k is a real number.) Rule 1: $log_a(M.N) = log_aM + log_aN$ Rule 2: $log_a \frac{M}{N} = log_aM - log_aN$ Rule 3: $log_a(M)^k = klog_aM$ Rule 4: $log_aa = 1$ Rule 5: $log_a 1 = 0$ Rule 6: $a^{log_ak} = k$
Example	**Evaluate** $3log_2(8)$ **Solution:** $8 = 2^3$, then $log_2(8) = log_2(2)^3$ Use log rule: $log_a(M)^k = klog_a(M) \rightarrow log_2(2)^3 = 3log_2(2)$ Use log rule: $log_a(a) = 1 \rightarrow 3 \times 3log_2{}^2 = 3 \times 3 = 9$
Your Turn!	1) $2log_3(27) = 6$ $\qquad$ 2) $3log_4(256) = 12$ 3) $\frac{1}{2}log_3(9) = 1$ $\qquad$ 4) $3log_5(25) = 6$ 5) $6log_3(3) = 6$ $\qquad$ 6) $4log_6(1) = 0$

Name:	Date: ....

Topic	**Natural Logarithms**
Notes	✓ A natural logarithm is a logarithm that has a special base of the mathematical constant e, which is an irrational number approximately equal to 2.71. ✓ The natural logarithm of x is generally written as $ln\ x$, or $log_e\ x$.
Example	**Solve this equation for x: $ln(3x - 4) = 1$** **Solution:** Use log rule: $a = log_b(b^a) \rightarrow 1 = ln(e^1) = ln(e) \rightarrow ln(3x - 4) = ln\ (e)$ When the logs have the same base: $log_b\big(f(x)\big) = log_b\big(g(x)\big) \rightarrow f(x) = g(x)$ $ln(3x - 4) = ln(e)$, then: $3x - 4 = e \rightarrow x = \frac{e+4}{3}$

Your Turn!	1) Solve for x: $e^x = 36$, $x = $ _____	2) Solve for x: $ln\ x = 5, x = $ _____
	3) Solve for x: $ln(2x - 3) = 1, x$ $= $ _____	4) Solve for x: $ln(ln\ x) = 2, x = $ _____
	5) Reduce this expressions to simplest form: $e^{ln\left(\frac{6}{e}\right)} = $ _____	6) Reduce this expressions to simplest form: $ln\left(\frac{1}{e}\right)^3 = $ _____

Name:	**Date:** ....

Topic	**Natural Logarithms - Answers**
Notes	✓ A natural logarithm is a logarithm that has a special base of the mathematical constant e, which is an irrational number approximately equal to 2.71. ✓ The natural logarithm of x is generally written as $ln\ x$, or $log_e\ x$.
Example	**Solve this equation for x: $ln(3x - 4) = 1$** **Solution:** Use log rule: $a = log_b(b^a) \rightarrow 1 = ln(e^1) = ln(e) \rightarrow ln(3x - 4) = ln\ (e)$ When the logs have the same base: $log_b\big(f(x)\big) = log_b\big(g(x)\big) \rightarrow f(x) = g(x)$ $ln(3x - 4) = ln(e)$, then: $3x - 4 = e \rightarrow x = \dfrac{e+4}{3}$
Your Turn!	1) Solve for x: $e^x = 36\ ,x = 2ln6$ 2) Solve for x: $ln\ x = 5, x = e^5$ 3) Solve for x: $ln(2x - 3) = 1, x = \dfrac{e+3}{2}$ 4) Solve for x: $ln(ln\ x) = 2, x = e^{e^2}$ 5) Reduce this expressions to simplest form: $e^{ln\left(\frac{6}{e}\right)} = \dfrac{6}{e}$ 6) Reduce this expressions to simplest form: $ln\left(\dfrac{1}{e}\right)^3 = -3$

Name:	Date: ....

Topic	Solving Logarithmic Equations
Notes	✓ Convert the logarithmic equation to an exponential equation when it's possible. (If no base is indicated, the base of the logarithm is 10) ✓ Condense logarithms if you have more than one log on one side of the equation. ✓ Plug in the answers back into the original equation and check to see if the solution works.
Example	**Find the value of the variables in this equation.** $$log_2(25 - x^2) = 4$$ **Solution:** Use the logarithmic definition: $log_a(b) = c \rightarrow a^c = b$ $$log_2(25 - x^2) = 4 \rightarrow 2^4 = (25 - x^2) \rightarrow 16 = (25 - x^2)$$ Simplify: $16 = (25 - x^2) \rightarrow -x^2 + 25 - 16 = 0$ Then: $x^2 = 9 \rightarrow x = 3 \ or -3$ Both 3 and -3 work in the original equation.

Your Turn!	1) Find the value of x: $log_3 4x = 0 \, , x = $ _____	2) Find the value of x: $logx + 4 = 1 \, , x = $ _____
	3) Find the value of x: $log3 - logx = 0 \, , x = $ _____	4) Find the value of x: $log(x - 3) - log6 = 0 \, , x = $ _____

Name:	Date: ....

Topic	Solving Logarithmic Equations - Answers
Notes	✔ Convert the logarithmic equation to an exponential equation when it's possible. (If no base is indicated, the base of the logarithm is 10) ✔ Condense logarithms if you have more than one log on one side of the equation. ✔ Plug in the answers back into the original equation and check to see if the solution works.
Example	**Find the value of the variables in this equation.** $$log_2(25 - x^2) = 4$$ **Solution:** Use the logarithmic definition: $log_a(b) = c \rightarrow a^c = b$ $$log_2(25 - x^2) = 4 \rightarrow 2^4 = (25 - x^2) \rightarrow 16 = (25 - x^2)$$ Simplify: $16 = (25 - x^2) \rightarrow -x^2 + 25 - 16 = 0$ Then: $x^2 = 9 \rightarrow x = 3 \ or -3$ Both 3 and -3 work in the original equation.

Your Turn!	1) Find the value of x: $log_3 4x = 0 , x = \dfrac{1}{4}$	2) Find the value of x: $logx + 4 = 1 , x = \dfrac{1}{1,000}$
	3) Find the value of x: $log3 - logx = 0 , x = 3$	4) Find the value of x: $log(x - 3) - log6 = 0 , x = 9$

Name:	Date: ....

Topic	**Circumference and Area of Circles**
Notes	✓ In a circle, variable r is usually used for the radius and d for diameter and π is about 3.14. ✓ *Area of a circle* $= \pi r^2$ ✓ *Circumference of a circle* $= 2\pi r$

Example	**Find the area of the circle.**
	  Solution:
	  Use area formula: $Area = \pi r^2$
	  $r = 2\ in \rightarrow Area = \pi(2)^2 = 4\pi, \pi = 3.14$
	  Then: $Area = 4 \times 3.14 = 12.56\ in^2$

	Find the area of each circle. ($\pi = 3.14$)

1) _____	2) _____
6 cm	10 in

Find the Circumference of each circle. ($\pi = 3.14$)

3) _____	4) _____
8 cm	6 m

Your Turn!

Name:	**Date:** ....

Topic	Circumference and Area of Circles - Answers
Notes	✓ In a circle, variable r is usually used for the radius and d for diameter and π is about 3.14. ✓ *Area of a circle* $= \pi r^2$ ✓ *Circumference of a circle* $= 2\pi r$

Example	**Find the area of the circle.**
	  Solution:
	  Use area formula: $Area = \pi r^2$
	  $r = 2\ in \rightarrow Area = \pi(2)^2 = 4\pi,\ \pi = 3.14$
	  Then: $Area = 4 \times 3.14 = 12.56\ in^2$

Your Turn!	**Find the area of each circle. ($\pi = 3.14$)** 1) $113.04\ cm^2$ 2) $314\ in^2$ **Find the Circumference of each circle. ($\pi = 3.14$)** 3) $50.24\ cm$ 4) $37.68\ m$

Name:	Date: ....

Topic	Arc Length and Sector Area
Notes	✓ To find the area of a sector of a circle, use this formula: Area of a sector $= \pi r^2 (\frac{\theta}{360})$, r is the radius of the circle and θ is the central angle of the sector. ✓ To find the arc of a sector of a circle, use this formula: Arc of a sector $= (\frac{\theta}{180})\pi r$
Examples	1) **Find the length of the arc. Round your answer to the nearest tenth.** $(\pi = 3.14),\ r = 8\ cm, \theta = 30°$ **Solution:** Use this formula: length of a sector $= (\frac{\theta}{180})\pi r$ Length of a sector $= (\frac{30}{180})\pi(8) = (\frac{1}{6})\pi(8) = 1.3 \times 3.14 \approx 4.2\ cm$ 2) **Find the area of the sector.** $r = 4\ ft, \theta = 80°$ **Solution:** Use this formula: Area of a sector $= \pi r^2(\frac{\theta}{360})$ Area of a sector $= \pi r^2 (\frac{\theta}{360}) = (3.14)(4^2)(\frac{80}{360}) \approx 11.2$
Your Turn!	**Find the length of the arc. Round your answer to the nearest tenth.** $(\pi = 3.14)$ <table><tr><td>1) $r = 12\ ft, \theta = 120°$ Length of the arc: _____</td><td>2) $r = 8\ cm, \theta = 32°$ Length of the arc: _____</td></tr></table>**Find the area of the sector. Round your answer to the nearest tenth.** $(\pi = 3.14)$ <table><tr><td>3) Find Area of the sector 16 in 320° Area of the sector: _____</td><td>4) Find Area of the sector $\frac{16\pi}{7}$ 12 ft Area of the sector: _____</td></tr></table>

| Name: .... | Date: .... |

Topic	**Arc Length and Sector Area - Answers**
Notes	✓ To find the area of a sector of a circle, use this formula: Area of a sector $= \pi r^2 (\frac{\theta}{360})$, r is the radius of the circle and θ is the central angle of the sector. ✓ To find the arc of a sector of a circle, use this formula: Arc of a sector $= (\frac{\theta}{180})\pi r$
Examples	1) **Find the length of the arc. Round your answer to the nearest tenth.** $(\pi = 3.14)$, $r = 8\ cm, \theta = 30°$ **Solution:** Use this formula: length of a sector $= (\frac{\theta}{180})\pi r$ Length of a sector $= (\frac{30}{180})\pi(8) = (\frac{1}{6})\pi(8) = 1.3 \times 3.14 \approx 4.2\ cm$ 2) **Find the area of the sector.** $r = 4\ ft, \theta = 80°$ **Solution:** Use this formula: Area of a sector $= \pi r^2 (\frac{\theta}{360})$ Area of a sector $= \pi r^2 (\frac{\theta}{360}) = (3.14)(4^2)(\frac{80}{360}) \approx 11.2$
Your Turn!	**Find the length of the arc. Round your answer to the nearest tenth. $(\pi = 3.14)$** 1) $r = 12\ ft, \theta = 120°$ 2) $r = 8\ cm, \theta = 32°$ Length of the arc: $25.1\ cm$ Length of the arc: $4.5\ cm$ **Find the area of the sector. Round your answer to the nearest tenth. $(\pi = 3.14)$** 3) Find Area of the sector. 4) Find Area of the sector. 16 in 320° $\frac{12\pi}{7}$ 12 ft Area of a sector: $714.5\ in^2$ Area of a sector: $387.6\ ft^2$

| Name: .... | Date: .... |

Topic	Equation of a Circle
Notes	✓ Equation of circles in standard form: $(x - h)^2 + (y - k)^2 = r^2$, Center: (h, k), Radius: r ✓ General format: $x^2 + y^2 + Ax + By + C = 0$

Example

Write the standard form equation of this circle.

$$x^2 + y^2 + 6x - 10y - 7 = 0$$

Solution: $(x - h)^2 + (y - k)^2 = r^2$ is the circle equation with a radius r, centered at (h, k). To find this equation, first, move the loose number to the right side: $x^2 + y^2 + 6x - 10y = 7$

Group x-variables and y-variables together:

$(x^2 + 6x) + (y^2 - 10y) = 7$

Convert x to square form:

$(x^2 + 6x + 9) + (y^2 - 10y) = 7 + 9 \rightarrow (x + 3)^2 + (y^2 - 10y) = 7 + 9 \rightarrow$
Convert y to square form:

$(x + 3)^2 + (y^2 - 10y + 25) = 7 + 9 + 25 \rightarrow$

$(x + 3)^2 + (y - 5)^2 = 41$

Then: $(x - (-3))^2 + (y - 5)^2 = \left(\sqrt{41}\right)^2$

Your Turn!

Write the standard form equation of each circle.

1) $x^2 + y^2 - 8x + 8y + 7 = 0$ Standard form: _____	2) $x^2 + y^2 - 4x + 10y + 13 = 0$ Standard form: _____
3) Center: $(-1, -2)$, Radius: 2 Standard form: _____	4) Center: $(-4, -1)$, Area: 4π Standard form: _____

| Name: .... | Date: .... |

Topic	**Equation of a Circle - Answers**
Notes	✓ Equation of circles in standard form: $(x - h)^2 + (y - k)^2 = r^2$, Center: (h, k), Radius: r ✓ General format: $x^2 + y^2 + Ax + By + C = 0$

	Write the standard form equation of this circle.
Example	$$x^2 + y^2 + 6x - 10y - 7 = 0$$ **Solution:** $(x - h)^2 + (y - k)^2 = r^2$ is the circle equation with a radius r, centered at (h, k). To find this equation, first, move the loose number to the right side: $x^2 + y^2 + 6x - 10y = 7$ Group x-variables and y-variables together: $(x^2 + 6x) + (y^2 - 10y) = 7$ Convert x to square form: $(x^2 + 6x + 9) + (y^2 - 10y) = 7 + 9 \rightarrow (x + 3)^2 + (y^2 - 10y) = 7 + 9 \rightarrow$ Convert y to square form: $(x + 3)^2 + (y^2 - 10y + 25) = 7 + 9 + 25 \rightarrow$ $(x + 3)^2 + (y - 5)^2 = 41$ Then: $(x - (-3))^2 + (y - 5)^2 = \left(\sqrt{41}\right)^2$

	Write the standard form equation of each circle.	
Your Turn!	1) $x^2 + y^2 - 8x + 8y + 7 = 0$ Standard form: $(x - 4)^2 + (y - (-4))^2 = 5^2$	2) $x^2 + y^2 - 4x + 10y + 13 = 0$ Standard form: $(x - 2)^2 + (y - (-5))^2 = 4^2$
	3) Center: $(-1, -2)$, Radius: 2 Standard form: $(x - (-1))^2 + (y - (-2))^2 = 2^2$	4) Center: $(-4, -1)$, Area: 4π Standard form: $(x - (-4))^2 + (y - (-1))^2 = 2^2$

Name:	Date: ....

Topic	**Finding the Center and the Radius of Circles**
Notes	To find the center and the radius of a circle using the equation of the circle: ✓ Write the equation of the circle in standard form: $(x - h)^2 + (y - k)^2 = r^2$, ✓ The center of the circle is at (h, k), and its radius is r.
Example	**Identify the center and radius.** $8x + x^2 - 2y = 8 - y^2$ **Solution:** $(x - h)^2 + (y - k)^2 = r^2$ is the circle equation with a radius r, centered at (h, k). Rewrite $8x + x^2 - 2y = 8 - y^2$ in the standard form: $$(x - (-4))^2 + (y - 1)^2 = 5^2$$ Then, the center is at $(-4, 1)$ and $r = 5$

Identify the center and radius of each circle.

1) $(x - 2)^2 + (y + 6)^2 = 16$ Center: (___,___) Radius: _____	2) $(x + 9)^2 + (y + 3)^2 = 20$ Center: (___,___) Radius: _____
3) $x^2 + y^2 + 12y = 3 + 10x$ Center: (___,___) Radius: _____	4) $x^2 - 8x + 16y = 1 - y^2$ Center: (___,___) Radius: _____

Your Turn!

Name: ....   **Date:** ....

Topic	Finding the Center and the Radius of Circles - Answers
Notes	To find the center and the radius of a circle using the equation of the circle: ✓ Write the equation of the circle in standard form: $(x - h)^2 + (y - k)^2 = r^2$, ✓ The center of the circle is at (h, k), and its radius is r. (h,k) r
Example	**Identify the center and radius.** $8x + x^2 - 2y = 8 - y^2$ **Solution:** $(x - h)^2 + (y - k)^2 = r^2$ is the circle equation with a radius r, centered at (h, k). Rewrite $8x + x^2 - 2y = 8 - y^2$ in the standard form: $$(x - (-4))^2 + (y - 1)^2 = 5^2$$ Then, the center is at $(-4, 1)$ and $r = 5$
Your Turn!	**Identify the center and radius of each circle.**

1) $(x - 2)^2 + (y + 6)^2 = 16$ Center: $(2, -6)$ Radius: 4	2) $(x + 9)^2 + (y + 3)^2 = 20$ Center: $(-9, -3)$ Radius: $\sqrt{20}$
3) $x^2 + y^2 + 12y = 3 + 10x$ Center: $(5, -6)$ Radius: 8	4) $x^2 - 8x + 16y = 1 - y^2$ Center: $(4, -8)$ Radius: 9

Name:	Date: ....

Topic	**Simplify Complex Fractions**
Notes	✓ Convert mixed numbers to improper fractions. ✓ Simplify all fractions. ✓ Write the fraction in the numerator of the main fraction line then write division sing (÷) and the fraction of the denominator. ✓ Use normal method for dividing fractions. ✓ Simplify as needed.
Example	**Solve:** $\dfrac{\frac{2}{3}}{\frac{7}{10} - \frac{1}{4}}$ **Solution:** First, simplify the denominator: $\dfrac{7}{10} - \dfrac{1}{4} = \dfrac{9}{20}$, Then: $\dfrac{\frac{2}{3}}{\frac{7}{10} - \frac{1}{4}} = \dfrac{\frac{2}{3}}{\frac{9}{20}}$ Now, write the complex fraction using the division sign (÷): $\dfrac{\frac{2}{3}}{\frac{9}{20}} = \dfrac{2}{3} \div \dfrac{9}{20}$ Use the dividing fractions rule: Keep, Change, Flip (keep the first fraction, change the division sign to multiplication, flip the second fraction) $\dfrac{2}{3} \div \dfrac{9}{20} = \dfrac{2}{3} \times \dfrac{20}{9} = \dfrac{40}{27} = 1\dfrac{13}{27}$
Your Turn!	1) $\dfrac{\frac{8}{3}}{\frac{2}{5}} = $ _____ 2) $\dfrac{\frac{x}{3} + \frac{x}{8}}{\frac{1}{4}} = $ _____ 3) $\dfrac{\frac{x+3}{3}}{\frac{x-2}{2}} = $ _____ 4) $\dfrac{1 + \frac{x}{4}}{x} = $ _____

Name:	Date: ....

Topic	Simplify Complex Fractions - Answers
Notes	✓ Convert mixed numbers to improper fractions. ✓ Simplify all fractions. ✓ Write the fraction in the numerator of the main fraction line then write division sing (÷) and the fraction of the denominator. ✓ Use normal method for dividing fractions. ✓ Simplify as needed.

Example

Solve: $\dfrac{\frac{2}{3}}{\frac{7}{10}-\frac{1}{4}}$

Solution: First, simplify the denominator: $\dfrac{7}{10}-\dfrac{1}{4}=\dfrac{9}{20}$, Then: $\dfrac{\frac{2}{3}}{\frac{7}{10}-\frac{1}{4}}=\dfrac{\frac{2}{3}}{\frac{9}{20}}$

Now, write the complex fraction using the division sign (÷):

$\dfrac{\frac{2}{3}}{\frac{9}{20}}=\dfrac{2}{3}\div\dfrac{9}{20}$

Use the dividing fractions rule: Keep, Change, Flip (keep the first fraction, change the division sign to multiplication, flip the second fraction)

$\dfrac{2}{3}\div\dfrac{9}{20}=\dfrac{2}{3}\times\dfrac{20}{9}=\dfrac{40}{27}=1\dfrac{13}{27}$

Your Turn!

1) $\dfrac{\frac{8}{3}}{\frac{2}{5}}=\dfrac{20}{3}$

2) $\dfrac{\frac{x}{3}+\frac{x}{8}}{\frac{1}{4}}=\dfrac{11x}{6}$

3) $\dfrac{\frac{x+3}{3}}{\frac{x-2}{2}}=\dfrac{2x+6}{3x-6}$

4) $\dfrac{1+\frac{x}{4}}{x}=\dfrac{4+x}{4x}$

Name:	Date: ....

Topic	**Graphing Rational Expressions**
Notes	✓ Find the vertical asymptotes of the function, if there is any. (Vertical asymptotes are vertical lines which correspond to the zeroes of the denominator) ✓ Find horizontal or slant asymptote. (If numerator has a bigger degree than denominator, there will be slant asymptote.) ✓ If denominator has a bigger degree than numerator, the horizontal asymptote is the x-axes or the line $y = 0$. If they have the same degree, the horizontal asymptote equals the leading coefficient (the coefficient of the largest exponent) of the numerator divided by the leading coefficient of the denominator. ✓ Find intercepts and plug in some values of x and solve for y and graph.
Example	**Graph rational expressions.** $f(x) = \dfrac{3x}{x^2-2x}$ **Solution:** First, notice that the graph is in two pieces. Find $y - intercept$ by substituting zero for x and solving for y $(f(x))$: $x = 0 \rightarrow y = \dfrac{3x}{x^2-2x} = \dfrac{3(0)}{0^2-2(0)} = \dfrac{0}{0}$, $y - intercept: None$ Asymptotes of $\dfrac{3x}{x^2-2x}$: vertical: $x = 2$, Horizontal: $y = 0$ After finding the asymptotes, you can plug in some values for x and solve for y. Here is the sketch for this function.
Your Turn!	1) Graph rational expressions. $f(x) = \dfrac{x^2-2x}{x-3}$ 2) Graph rational expressions. $f(x) = \dfrac{6x+1}{x^2-4x}$ 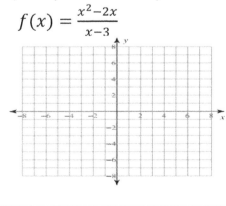

Name:	Date: ....

Topic	**Graphing Rational Expressions**
Notes	✓ Find the vertical asymptotes of the function, if there is any. (Vertical asymptotes are vertical lines which correspond to the zeroes of the denominator) ✓ Find horizontal or slant asymptote. (If numerator has a bigger degree than denominator, there will be slant asymptote.) ✓ If denominator has a bigger degree than numerator, the horizontal asymptote is the x-axes or the line $y = 0$. If they have the same degree, the horizontal asymptote equals the leading coefficient (the coefficient of the largest exponent) of the numerator divided by the leading coefficient of the denominator. ✓ Find intercepts and plug in some values of x and solve for y and graph.
Example	**Graph rational expressions.** $f(x) = \frac{3x}{x^2 - 2x}$ **Solution:** First, notice that the graph is in two pieces. Find $y - intercept$ by substituting zero for x and solving for y ($f(x)$): $x = 0 \rightarrow y = \frac{3x}{x^2 - 2x} = \frac{3(0)}{0^2 - 2(0)} = \frac{0}{0}$ $y - intercept$: $None$ Asymptotes of $\frac{3x}{x^2 - 2x}$: vertical: $x = 2$, Horizontal: $y = 0$ After finding the asymptotes, you can plug in some values for x and solve for y. Here is the sketch for this function.
Your Turn!	1) Graph rational expressions. $f(x) = \frac{x^2 - 2x}{x - 3}$ 2) Graph rational expressions. $f(x) = \frac{6x + 1}{x^2 - 4x}$

Name:	**Date:** ....

Topic	**Adding and Subtracting Rational Expressions**
Notes	For adding and subtracting rational expressions: ✓ Find least common denominator (LCD). ✓ Write each expression using the LCD. ✓ Add or subtract the numerators. ✓ Simplify as needed.
Examples	1) Solve. $\dfrac{3}{x+4} + \dfrac{x-2}{x+4} =$ Use fraction addition rule: $\dfrac{a}{c} \pm \dfrac{b}{c} = \dfrac{a \pm b}{c} \rightarrow \dfrac{3}{x+4} + \dfrac{x-2}{x+4} = \dfrac{3+(x-2)}{x+4} =$ $\dfrac{x+1}{x+4}$ 2) Solve. $\dfrac{x+4}{x-8} + \dfrac{x}{x+6} =$ Least common denominator of $(x-8)$ and $(x+6)$: $(x-8)(x+6)$ Then: $\dfrac{(x+4)(x+6)}{(x-8)(x+6)} + \dfrac{x(x-8)}{(x+6)(x-8)} = \dfrac{(x+4)(x+6)+x(x-8)}{(x+6)(x-8)}$ Expand: $(x+4)(x+6) + x(x-8) = 2x^2 + 2x + 24$ Then: $\dfrac{x+4}{x-8} + \dfrac{x}{x+6} = \dfrac{2x^2+2x+24}{(x+6)(x-8)}$
Your Turn!	1) $\dfrac{x+6}{x+1} - \dfrac{x+9}{x+1} =$ _____ 2) $\dfrac{2x+1}{x+3} + \dfrac{2}{x+4} =$ _____ 3) $\dfrac{14}{x+4} + \dfrac{6}{x^2-16} =$ _____ 4) $\dfrac{x+2}{x+8} - \dfrac{2x}{x-8} =$ _____

Name: ....   Date: ....

Topic	**Adding and Subtracting Rational Expressions - Answers**
Notes	For adding and subtracting rational expressions: ✓ Find least common denominator (LCD). ✓ Write each expression using the LCD. ✓ Add or subtract the numerators. ✓ Simplify as needed.
Examples	1) Solve. $\dfrac{3}{x+4} + \dfrac{x-2}{x+4} =$ Use fraction addition rule: $\dfrac{a}{c} \pm \dfrac{b}{c} = \dfrac{a \pm b}{c} \rightarrow \dfrac{3}{x+4} + \dfrac{x-2}{x+4} = \dfrac{3+(x-2)}{x+4} =$ $\dfrac{x+1}{x+4}$ 2) Solve. $\dfrac{x+4}{x-8} + \dfrac{x}{x+6} =$ Least common denominator of $(x-8)$ and $(x+6)$: $(x-8)(x+6)$ Then: $\dfrac{(x+4)(x+6)}{(x-8)(x+6)} + \dfrac{x(x-8)}{(x+6)(x-8)} = \dfrac{(x+4)(x+6)+x(x-8)}{(x+6)(x-8)}$ Expand: $(x+4)(x+6) + x(x-8) = 2x^2 + 2x + 24$ Then: $\dfrac{x+4}{x-8} + \dfrac{x}{x+6} = \dfrac{2x^2+2x+24}{(x+6)(x-8)}$

Your Turn!

1) $\dfrac{x+6}{x+1} - \dfrac{x+9}{x+1} = -\dfrac{3}{x+1}$

2) $\dfrac{2x+1}{x+3} + \dfrac{2}{x+4} = \dfrac{2x^2+11x+10}{(x+3)(x+4)}$

3) $\dfrac{14}{x+4} + \dfrac{6}{x^2-16} = \dfrac{14x-50}{(x+4)(x-4)}$

4) $\dfrac{x+2}{x+8} - \dfrac{2x}{x-8} = \dfrac{-x^2-22x-16}{(x+8)(x-8)}$

Name:	Date: ....

Topic	**Multiplying Rational Expressions**
Notes	✓ Multiplying rational expressions is the same as multiplying fractions. First, multiply numerators and then multiply denominators. Then, simplify as needed.
Examples	**1)** Solve: $\frac{x+5}{x-1} \times \frac{x-1}{3} =$ Multiply fractions: $\frac{x+5}{x-1} \times \frac{x-1}{3} = \frac{(x+5)(x-1)}{3(x-1)}$ Cancel the common factor: $(x-1)$, then: $\frac{(x+5)(x-1)}{3(x-1)} = \frac{(x+5)}{3}$ **2)** Solve: $\frac{x-5}{x+4} \times \frac{2x+8}{x-5} =$ Multiply fractions: $\frac{x-5}{x+4} \times \frac{2x+8}{x-5} = \frac{(x-5)(2x+8)}{(x+4)(x-5)}$ Cancel the common factor: $\frac{(x-5)(2x+8)}{(x+4)(x-5)} = \frac{(2x+8)}{(x+4)}$ Factor $2x + 8 = 2(x + 4)$, Then: $\frac{2(x+4)}{(x+4)} = 2$
Your Turn!	1) $\frac{20x^3}{3} \times \frac{15}{4x} = $ _____ 2) $\frac{x+6}{4} \times \frac{16}{x+6} = $ _____ 3) $\frac{x+10}{4x} \times \frac{3x}{7x+70} = $ _____ 4) $\frac{x+8}{x+6} \times \frac{x-6}{4x+32} = $ _____

Name:	Date: ....

Topic	**Multiplying Rational Expressions - Answers**
Notes	✓ Multiplying rational expressions is the same as multiplying fractions. First, multiply numerators and then multiply denominators. Then, simplify as needed.
Examples	**1) Solve:** $\frac{x+5}{x-1} \times \frac{x-1}{3} =$ Multiply fractions: $\frac{x+5}{x-1} \times \frac{x-1}{3} = \frac{(x+5)(x-1)}{3(x-1)}$ Cancel the common factor: $(x-1)$, then: $\frac{(x+5)(x-1)}{3(x-1)} = \frac{(x+5)}{3}$ **2) Solve:** $\frac{x-5}{x+4} \times \frac{2x+8}{x-5} =$ Multiply fractions: $\frac{x-5}{x+4} \times \frac{2x+8}{x-5} = \frac{(x-5)(2x+8)}{(x+4)(x-5)}$ Cancel the common factor: $\frac{(x-5)(2x+8)}{(x+4)(x-5)} = \frac{(2x+8)}{(x+4)}$ Factor $2x+8 = 2(x+4)$, Then: $\frac{2(x+4)}{(x+4)} = 2$
Your Turn!	1) $\frac{20x^3}{3} \times \frac{15}{4x} = 25x^2$ 2) $\frac{x+6}{4} \times \frac{16}{x+6} = 4$ 3) $\frac{x+10}{4x} \times \frac{3x}{7x+70} = \frac{3}{28}$ 4) $\frac{x+8}{x+6} \times \frac{x-6}{4x+32} = \frac{x-6}{4(x+6)}$

| Name: .... | Date: .... |

Topic	**Dividing Rational Expressions**
Notes	✓ To divide rational expression, use the same method we use for dividing fractions. ✓ Keep, Change, Flip ✓ Keep first rational expression, change division sign to multiplication, and flip the numerator and denominator of the second rational expression. Then, multiply numerators and multiply denominators. Simplify as needed.
Examples	1) Solve $\dfrac{2x}{5} \div \dfrac{8}{7} =$ $\dfrac{2x}{5} \div \dfrac{8}{7} = \dfrac{\frac{2x}{5}}{\frac{8}{7}}$, Use Divide fractions rules: $\dfrac{\frac{a}{b}}{\frac{c}{d}} = \dfrac{a.d}{b.c}$ $\dfrac{\frac{2x}{5}}{\frac{8}{7}} = \dfrac{2x \times 7}{8 \times 5} = \dfrac{14x}{40} = \dfrac{7x}{20}$ 2) Solve $\dfrac{6x}{x+2} \div \dfrac{x}{6x+12} =$ $\dfrac{\frac{6x}{x+2}}{\frac{x}{6x+12}}$, Use Divide fractions rules: $\dfrac{(6x)(6x+12)}{(x)(x+2)}$ Cancel common fraction: $\dfrac{(6x)(6x+12)}{(x)(x+2)} = \dfrac{36(x+2)}{(x+2)} = 36$
Your Turn!	1) $\dfrac{10x}{x+2} \div \dfrac{x}{60x+120} =$ _____ 2) $\dfrac{5}{4} \div \dfrac{45}{8x} =$ _____ 3) $\dfrac{x-6}{x+3} \div \dfrac{4}{x+3} =$ _____ 4) $\dfrac{7x^3}{x^2-64} \div \dfrac{x^3}{x^2+x-56} =$ _____

Name:	Date: ....

Topic	**Dividing Rational Expressions - Answers**
Notes	✓ To divide rational expression, use the same method we use for dividing fractions. ✓ Keep, Change, Flip ✓ Keep first rational expression, change division sign to multiplication, and flip the numerator and denominator of the second rational expression. Then, multiply numerators and multiply denominators. Simplify as needed.
Examples	**1)** Solve $\dfrac{2x}{5} \div \dfrac{8}{7} =$ $\dfrac{2x}{5} \div \dfrac{8}{7} = \dfrac{\frac{2x}{5}}{\frac{8}{7}}$, Use Divide fractions rules: $\dfrac{\frac{a}{b}}{\frac{c}{d}} = \dfrac{a.d}{b.c}$ $\dfrac{\frac{2x}{5}}{\frac{8}{7}} = \dfrac{2x \times 7}{8 \times 5} = \dfrac{14x}{40} = \dfrac{7x}{20}$ **2)** Solve $\dfrac{6x}{x+2} \div \dfrac{x}{6x+12} =$ $\dfrac{\frac{6x}{x+2}}{\frac{x}{6x+12}}$, Use Divide fractions rules: $\dfrac{(6x)(6x+12)}{(x)(x+2)}$ Cancel common fraction: $\dfrac{(6x)(6x+12)}{(x)(x+2)} = \dfrac{36(x+2)}{(x+2)} = 36$
Your Turn!	**1)** $\dfrac{10x}{x+2} \div \dfrac{x}{60x+120} =$ 600 **2)** $\dfrac{5}{4} \div \dfrac{45}{8x} =$ $\dfrac{2x}{9}$ **3)** $\dfrac{x-6}{x+3} \div \dfrac{4}{x+3} =$ $\dfrac{x-6}{4}$ **4)** $\dfrac{7x^3}{x^2-64} \div \dfrac{x^3}{x^2+x-56} =$ $\dfrac{7(x-7)}{x-8}$

Name:	Date: ....

Topic	Rational Equations
Notes	For solving rational equations, we can use following methods: ✓ Converting to a common denominator: In this method, you need to get a common denominator for both sides of the equation. Then make the numerators equal and solve for the variable. ✓ Cross-multiplying: This method is useful when there is only one fraction on each side of the equation. Simply multiply the first numerator by the second denominator and make the result equal to the product of the second numerator and the first denominator.
Example	**Solve.** $\frac{x-3}{x+1} = \frac{x+5}{x-2}$ Use cross multiply method: if $\frac{a}{b} = \frac{c}{d}$, then: $a \times d = b \times c$ Then: $(x-3)(x-2) = (x+5)(x+1)$ Expand: $(x-3)(x-2) = x^2 - 5x + 6$ Expand: $(x+5)(x+1) = x^2 + 6x + 5$, Then: $x^2 - 5x + 6 = x^2 + 6x + 5$, Simplify: $x^2 - 5x = x^2 + 6x - 1$ Subtract both sides $x^2 + 6x$,Then: $-11x = -1 \rightarrow x = \frac{1}{11}$
Your Turn!	1) $\frac{1}{x^2} + \frac{4}{x} = \frac{6}{x}$ $x = \_\_\_$ 2) $\frac{2}{x^2} - \frac{1}{x} = 1$ $x = \_\_\_$ 3) $\frac{x+1}{5x} - 1 = \frac{1}{x}$ $x = \_\_\_$ 4) $\frac{6}{x} - \frac{1}{x} = \frac{1}{x^2+6x}$ $x = \_\_\_$

Name:	Date: ....

Topic	**Rational Equations - Answers**	
Notes	For solving rational equations, we can use following methods: ✓ Converting to a common denominator: In this method, you need to get a common denominator for both sides of the equation. Then make the numerators equal and solve for the variable. ✓ Cross-multiplying: This method is useful when there is only one fraction on each side of the equation. Simply multiply the first numerator by the second denominator and make the result equal to the product of the second numerator and the first denominator.	
Example	**Solve.** $\frac{x-3}{x+1} = \frac{x+5}{x-2}$ Use cross multiply method: if $\frac{a}{b} = \frac{c}{d}$, then: $a \times d = b \times c$ Then: $(x-3)(x-2) = (x+5)(x+1)$ Expand: $(x-3)(x-2) = x^2 - 5x + 6$ Expand: $(x+5)(x+1) = x^2 + 6x + 5$, Then: $x^2 - 5x + 6 = x^2 + 6x + 5$, Simplify: $x^2 - 5x = x^2 + 6x - 1$ Subtract both sides $x^2 + 6x$,Then: $-11x = -1 \rightarrow x = \frac{1}{11}$	
Your Turn!	1) $\frac{1}{x^2} + \frac{4}{x} = \frac{6}{x}$ $x = \frac{1}{2}$	2) $\frac{2}{x^2} - \frac{1}{x} = 1$ $x = 1 \ or \ x = -2$
	3) $\frac{x+1}{5x} - 1 = \frac{1}{x}$ $x = -1$	4) $\frac{6}{x} - \frac{1}{x} = \frac{1}{x^2+6x}$ $x = -\frac{29}{5}$

Name:	Date: ....

Topic	Angle and Angle Measure
Notes	✓ To convert degrees to radians, use this formula: $\text{Radians} = \text{Degrees} \times \frac{\pi}{180}$ ✓ To convert radians to degrees, use this formula: $\text{Degrees} = \text{Radians} \times \frac{180}{\pi}$
Examples	1) **Convert 140 degrees to radians.** **Solution:** Use this formula: $\text{Radians} = \text{Degrees} \times \frac{\pi}{180}$ $\text{Radian} = 140 \times \frac{\pi}{180} = \frac{140\pi}{180} = \frac{7}{9}\pi$ 2) **Convert $\frac{9}{5}\pi$ to degrees.** **Solution:** Use this formula: $\text{Degrees} = \text{Radians} \times \frac{180}{\pi}$ $\text{Degree} = \frac{9\pi}{5} \times \frac{180}{\pi} = \frac{1,620\pi}{5\pi} = 324°$
Your Turn!	**Convert each degree measure into radians.** <table><tr><td>1) $50° =$</td><td>2) $-110° =$</td></tr></table> **Convert each radian measure into degrees.** <table><tr><td>3) $\frac{\pi}{5} =$</td><td>4) $-\frac{3\pi}{2} =$</td></tr></table>

Name: ....   Date: ....

Topic	Angle and Angle Measure - Answers
Notes	✓ To convert degrees to radians, use this formula: Radians = Degrees $\times \frac{\pi}{180}$ ✓ To convert radians to degrees, use this formula: Degrees = Radians $\times \frac{180}{\pi}$
Examples	1) **Convert 140 degrees to radians.** **Solution:** Use this formula: Radians = Degrees $\times \frac{\pi}{180}$ Radian = $140 \times \frac{\pi}{180} = \frac{140\pi}{180} = \frac{7}{9}\pi$ 2) **Convert $\frac{9}{5}\pi$ to degrees.** **Solution:** Use this formula: Degrees = Radians $\times \frac{180}{\pi}$ Degree = $\frac{9\pi}{5} \times \frac{180}{\pi} = \frac{1,620\pi}{5\pi} = 324°$
Your Turn!	**Convert each degree measure into radians.** 1) $50° = \frac{5}{18}\pi$ | 2) $-110° = -\frac{11}{18}\pi$ **Convert each radian measure into degrees.** 3) $\frac{\pi}{5} = 36°$ | 4) $-\frac{3\pi}{2} = -270°$

Name: ....   Date: ....

Topic	Trigonometric Functions

Notes

✓ Learn common trigonometric functions:

θ	0°	30°	45°	60°	90°
$\sin\theta$	0	$\dfrac{1}{2}$	$\dfrac{\sqrt{2}}{2}$	$\dfrac{\sqrt{3}}{2}$	1
$\cos\theta$	1	$\dfrac{\sqrt{3}}{2}$	$\dfrac{\sqrt{2}}{2}$	$\dfrac{1}{2}$	0
$\tan\theta$	0	$\dfrac{\sqrt{3}}{3}$	1	$\sqrt{3}$	Undefined

Example

Find sin 135°.

Solution: Use the following property: $sin(x) = cos\,(90° - x)$

$sin\,135° = cos\,(90° - 135°) = cos(-45°)$

Now use the following property: $cos(-x) = cos\,(x)$

$cos\,(-45°) = cos\,(45°) = \dfrac{\sqrt{2}}{2}$

Your Turn!

1) $sin\,(30°) =$ _____

2) $cot\,(30°) =$ _____

3) $sin(-90°) =$ _____

4) $cos\,(-60°\,) =$ _____

5) $tan(-120°) =$ _____

6) $sec\,(480°) =$ _____

Name: ....   **Date:** ....

Topic	**Trigonometric Functions- Answers**				
Notes	✓ Learn common trigonometric functions:				

θ	0°	30°	45°	60°	90°
$\sin\theta$	0	$\frac{1}{2}$	$\frac{\sqrt{2}}{2}$	$\frac{\sqrt{3}}{2}$	1
$\cos\theta$	1	$\frac{\sqrt{3}}{2}$	$\frac{\sqrt{2}}{2}$	$\frac{1}{2}$	0
$\tan\theta$	0	$\frac{\sqrt{3}}{3}$	1	$\sqrt{3}$	Undefined

Example

Find sin 135°.

Solution: Use the following property: $sin(x) = cos(90° - x)$

$sin\,135° = cos(90° - 135°) = cos(-45°)$

Now use the following property: $cos(-x) = cos\,(x)$

$cos(-45°) = cos(45°) = \frac{\sqrt{2}}{2}$

Your Turn!

1) $sin\,(30°) = \frac{1}{2}$

2) $cot\,(30°) = \sqrt{3}$

3) $sin(-90°) = -1$

4) $cos\,(-60°) = \frac{1}{2}$

5) $tan(-120°) = \sqrt{3}$

6) $sec\,(480°) = -2$

Name:	Date: ....

Topic	Coterminal Angles and Reference Angles
Notes	✓ Coterminal angles are equal angles. ✓ To find a Coterminal of an angle, add or subtract 360 degrees (or 2π for radians) to the given angle. ✓ Reference angle is the smallest angle that you can make from the terminal side of an angle with the x-axis.
Example	**Find a positive and a negative Coterminal angles to angle $95°$.** **Solution:** $95° - 360° = -265°$ $95° + 360° = 455°$ $-265°$ and a $455°$ are Coterminal with a $95°$.

Find a positive and a negative Coterminal angles.

Your Turn!	1) $80° =$ Positive = _____ Negative = _____	2) $120° =$ Positive = _____ Negative = _____
	3) $-115° =$ Positive = _____ Negative = _____	4) $\dfrac{\pi}{5} =$ Positive = _____ Negative = _____
	5) $\dfrac{3\pi}{4} =$ Positive = _____ Negative = _____	6) $\dfrac{2\pi}{7} =$ Positive = _____ Negative = _____

Name:	Date: ....

Topic	**Coterminal Angles and Reference Angles- Answers**
Notes	✓ Coterminal angles are equal angles. ✓ To find a Coterminal of an angle, add or subtract 360 degrees (or 2π for radians) to the given angle. ✓ Reference angle is the smallest angle that you can make from the terminal side of an angle with the x-axis.
Example	**Find a positive and a negative Coterminal angles to angle 95°.** **Solution:** $95° - 360° = -265°$ $95° + 360° = 455°$ $-265°$ and a $455°$ are Coterminal with a $95°$.

Find a positive and a negative Coterminal angles.

Your Turn!	1) $80° =$ Positive $= 440°$ Negative $= -280°$	2) $120° =$ Positive $= 480°$ Negative $= -240°$
	3) $-115° =$ Positive $= 245°$ Negative $= -475°$	4) $\dfrac{\pi}{5} =$ Positive $= \dfrac{11\pi}{5}$ Negative $= -\dfrac{9\pi}{5}$
	5) $\dfrac{3\pi}{4} =$ Positive $= \dfrac{11\pi}{4}$ Negative $= -\dfrac{5\pi}{4}$	6) $\dfrac{2\pi}{7} =$ Positive $= \dfrac{16\pi}{7}$ Negative $= -\dfrac{12\pi}{7}$

Name:	Date: ....

Topic	Evaluating Trigonometric Function
Notes	✓ Step 1: Find the reference angle. (It is the smallest angle that you can make from the terminal side of an angle with the x-axis.) ✓ Step 2: Determine the quadrant of the function. Depending on the quadrant in which the function lies, the answer will be either be positive or negative. ✓ Step 3: Find the trigonometric function of the reference angle.
Example	**Find the exact value of trigonometric function.** $\tan \frac{4\pi}{3}$ **Solution:** Rewrite the angles for $\tan \frac{4\pi}{3}$: $\tan \frac{4\pi}{3} = \tan \left(\frac{3\pi+\pi}{3}\right) = \tan \left(\pi + \frac{1}{3}\pi\right)$. Use the periodicity of $\tan$: $\tan(x + \pi . k) = \tan(x)$. Recall that $\tan \frac{1}{3}\pi = \sqrt{3}$ Then: $\tan \left(\pi + \frac{1}{3}\pi\right) = \tan \left(\frac{1}{3}\pi\right) = \sqrt{3}$
Your Turn!	1) $\cot 150° = $ _____ 2) $\sin 120° = $ _____ 3) $\tan 300° = $ _____ 4) $\cot \frac{5\pi}{3} = $ _____ 5) $\cos \frac{11\pi}{6} = $ _____ 6) $\csc \frac{5\pi}{6} = $ _____

Name:	Date: ....

Topic	**Evaluating Trigonometric Function- Answers**
Notes	✓ Step 1: Find the reference angle. (It is the smallest angle that you can make from the terminal side of an angle with the x-axis.) ✓ Step 2: Determine the quadrant of the function. Depending on the quadrant in which the function lies, the answer will be either be positive or negative. ✓ Step 3: Find the trigonometric function of the reference angle.
Example	**Find the exact value of trigonometric function.** $tan\ \frac{4\pi}{3}$ **Solution:** Rewrite the angles for $tan\ \frac{4\pi}{3}$: $tan\ \frac{4\pi}{3} = tan\left(\frac{3\pi+\pi}{3}\right) = tan\left(\pi + \frac{1}{3}\pi\right).$ Use the periodicity of tan: $tan(x + \pi . k) = tan(x).$ Recall that $tan\frac{1}{3}\pi = \sqrt{3}$ Then: $tan\left(\pi + \frac{1}{3}\pi\right) = tan\left(\frac{1}{3}\pi\right) = \sqrt{3}$
Your Turn!	1) $cot\ 150° = -\sqrt{3}$ 2) $sin\ 120° = \frac{\sqrt{3}}{2}$ 3) $tan\ 300° = -\sqrt{3}$ 4) $cot\ \frac{5\pi}{3} = -\frac{\sqrt{3}}{3}$ 5) $cos\ \frac{11\pi}{6} = \frac{\sqrt{3}}{2}$ 6) $csc\ \frac{5\pi}{6} = 2$

Time to Test

Time to refine your skill with a practice examination

Take a REAL ALEKS Mathematics test to simulate the test day experience. After you've finished, score your test using the answers and explanations section.

Before You Start

- ❖ You'll need a pencil and scratch papers to take the test.
- ❖ For these practice tests, don't time yourself. Spend time as much as you need.
- ❖ After you've finished the test, review the answer key to see where you went wrong.

Good luck!

ALEKS Mathematics

Practice Test 1

2020 - 2021

Total number of questions: 30

Total time (Calculator): No time limit

Calculators are permitted for ALEKS Math Test.

(On a real ALEKS test, there is an onscreen calculator to use.)

207

1) If $f(x) = 3x - 1$ and $g(x) = x^2 - x$, then find $\left(\frac{f}{g}\right)(x)$.

2) A bank is offering 3.5% simple interest on a savings account. If you deposit $12,000, how much interest will you earn in two years?

3) If the ratio of home fans to visiting fans in a crowd is $3:2$ and all 25,000 seats in a stadium are filled, how many visiting fans are in attendance?

4) If the interior angles of a quadrilateral are in the ratio $1:2:3:4$, what is the measure of the largest angle?

5) If the area of a circle is 64 square meters, what is its diameter?

6) The length of a rectangle is $\frac{5}{4}$ times its width. If the width is 16, what is the perimeter of this rectangle?

7) In the figure below, line A is parallel to line B. What is the value of angle x?

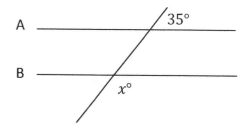

8) An angle is equal to one fifth of its supplement. What is the measure of that angle?

9) What is the value of x in the following system of equations?
$$2x + 5y = 11$$
$$4x - 2y = -14$$

10) Last week 24,000 fans attended a football match. This week three times as many bought tickets, but one sixth of them cancelled their tickets. How many are attending this week?

11) If $sin A = \frac{1}{4}$ in a right triangle and the angle A is an acute angle, then what is $cos A$?

12) In the standard (x, y) coordinate system plane, what is the area of the circle with the following equation?

$$(x + 2)^2 + (y - 4)^2 = 16$$

13) Convert 670,000 to scientific notation.

14) The ratio of boys to girls in a school is $2 : 3$. If there are 600 students in a school, how many boys are in the school?

15) If 150% of a number is 75, then what is 90% of that number?

16) If $A = \begin{bmatrix} -1 & 2 \\ 1 & -2 \end{bmatrix}$ and $B = \begin{bmatrix} 4 & 1 \\ -2 & 3 \end{bmatrix}$, then $2A - B =$

17) What is the solution of the following inequality?
$$|x - 2| \geq 3$$

18) If $\tan x = \frac{8}{15}$, then $\sin x =$

19) $\left(x^6\right)^{\frac{5}{8}}$ equal to?

20) What are the zeroes of the function $f(x) = x^3 + 6x^2 + 8x$?

21) If $x + sin^2 a + cos^2 a = 3$, then $x = ?$

22) If $\sqrt{6x} = \sqrt{y}$, then $x =$

23) The average weight of 18 girls in a class is 60 kg and the average weight of 32 boys in the same class is 62 kg. What is the average weight of all the 50 students in that class?

24) What is the value of the expression $5(x - 2y) + (2 - x)^2$ when $x = 3$ and $y = -2$?

25) Sophia purchased a sofa for \$530.40. The sofa is regularly priced at \$624. What was the percent discount Sophia received on the sofa?

26) If one angle of a right triangle measures 60°, what is the sine of the other acute angle?

27) Simplify $\dfrac{5 - 3i}{-5i}$?

28) The average of five consecutive numbers is 38. What is the smallest number?

29) What is the slope of a line that is perpendicular to the line?
$$4x - 2y = 12$$

30) If $f(x) = 2x^4 + 2$ and $(x) = \frac{1}{x}$, what is the value of $f(g(x))$?

This is the end of Practice Test 1.

ALEKS Mathematics

Practice Test 2

2020 - 2021

Total number of questions: 30

Total time (Calculator): No time limit

Calculators are permitted for ALEKS Math Test.

(On a real ALEKS test, there is an onscreen calculator to use.)

215

1) $(x - 5)(x^2 + 5x + 4) = ?$

2) $5 + 8 \times (-3) - [4 + 22 \times 5] \div 6 = ?$

3) Simplify. $\dfrac{\frac{1}{2} - \frac{x+5}{4}}{\frac{x^2}{2} - \frac{5}{2}}$

4) How many 4×2 squares can fit inside a rectangle with a height of 52 and width of 12?

5) If $5 + 2x \le 15$, what is the value of $x \le$?

6) A man owed $4,265 on his car. After making 55 payments of $66 each, how much did he have left to pay?

7) $(x^4)^{\frac{5}{8}} =$

8) What is 2531.58245 rounded to the nearest tenth?

9) 25 is what percent of 20?

10) Last Friday Jacob had $34.52. Over the weekend he received some money for cleaning the attic. He now has $44. How much money did he receive?

11) In the following triangle what is the value of x?

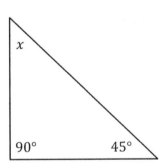

12) Find the factors of $x^2 - 7x + 12$.

13) A ladder leans against a wall forming a $60°$ angle between the ground and the ladder. If the bottom of the ladder is 30 feet away from the wall, how long is the ladder?

14) What is the distance between the points $(-2, 7)$ and $(1, 3)$?

15) Write the $\frac{2}{140}$ as a decimal. (round your answer to the nearest ten thousandths)

16) Liam's average (arithmetic mean) on two mathematics tests is 9. What should Liam's score be on the next test to have an overall of 10 for all the tests?

17) Find all values of x in this equation: $4x^2 + 14x + 6 = 0$

18) What is the value of x in this equation? $7^5 \times 7^8 = 7^x$

19) If a vehicle is driven 33 miles on Monday, 36 miles on Tuesday, and 30 miles on Wednesday, what is the average number of miles driven each day?

20) Find the solutions of the following equation.

$$x^2 + 2x - 5 = 0$$

21) What is the solution of the following system of equations?

$$\begin{cases} -2x - y = -9 \\ 5x - 2y = 9 \end{cases}$$

22) Solve.

$$|9 - (12 \div |2 - 6|)| = ?$$

23) If $\log_2 x = 5$, then $x = ?$

24) What's the reciprocal of $\dfrac{x^3}{14}$?

25) What is the equivalent temperature of $140°F$ in Celsius?

$$C = \frac{5}{9}(F - 32)$$

26) Simplify $(-3 + 9i)(3 + 5i)$.

27) Find $tan\frac{2\pi}{3}$

28) If $f(x) = 3x + 4(x + 1) + 2$ then $f(4x) = ?$

29) What is the center and radius of a circle with the following equation?

$$(x - 4)^2 + (y + 7)^2 = 3$$

30) If the center of a circle is at the point $(-4, 2)$ and its circumference equals to 2π, what is the standard form equation of the circle?

This is the end of Practice Test 2.

ALEKS Mathematics Practice Tests Answers and

Explanations

ALEKS Mathematics Practice Test 1

1) The answer is $\frac{3x-1}{x^2-x}$

$$\left(\frac{f}{g}\right)(x) = \frac{f(x)}{g(x)} = \frac{3x-1}{x^2-x}$$

2) The answer is 840

Use simple interest formula: $I = prt$ (I = interest, p = principal, r = rate, t = time)

$I = (12,000)(0.035)(2) = 840$

3) The answer is $10,000$

Number of visiting fans: $\frac{2 \times 25,000}{5} = 10,000$

4) The answer is $144°$

The sum of all angles in a quadrilateral is 360 degrees. Let x be the smallest angle in the quadrilateral. Then the angles are: $x, 2x, 3x, 4x$, $x + 2x + 3x + 4x = 360 \rightarrow 10x = 360 \rightarrow x = 36$, The angles in the quadrilateral are: $36°, 72°, 108°$, and $144°$

5) The answer is $\frac{8\sqrt{\pi}}{\pi}$

Formula for the area of a circle is: $A = \pi r^2$, Using 64 for the area of the circle we have:

$64 = \pi r^2$. Let's solve for the radius (r). $\frac{64}{\pi} = r^2 \rightarrow r = \sqrt{\frac{64}{\pi}} = \frac{8}{\sqrt{\pi}} = \frac{8}{\sqrt{\pi}} \times \frac{\sqrt{\pi}}{\sqrt{\pi}} = \frac{8\sqrt{\pi}}{\pi}$

6) The answer is 72

Length of the rectangle is: $\frac{5}{4} \times 16 = 20$, perimeter of rectangle is: $2 \times (20 + 16) = 72$

7) The answer is $145°$

The angle x and 35 are complementary angles. Therefore: $x + 35 = 180 \rightarrow$

$x = 180° - 35° = 145°$

8) The answer is 30

The sum of supplement angles is 180. Let x be that angle. Therefore, $x + 5x = 180$

$6x = 180$, divide both sides by 6: $x = 30$

9) The answer is -2

Solving Systems of Equations by Elimination: Multiply the first equation by (-2), then add it to the second equation.

$$\begin{array}{c} -2(2x + 5y = 11) \\ \underline{4x - 2y = -14} \end{array} \Rightarrow \begin{array}{c} -4x - 10y = -22 \\ 4x - 2y = -14 \end{array} \Rightarrow -12y = -36 \Rightarrow y = 3$$

Plug in the value of y into one of the equations and solve for x.

$2x + 5(3) = 11 \Rightarrow 2x + 15 = 11 \Rightarrow 2x = -4 \Rightarrow x = -2$

10) The answer is $60,000$

Three times of 24,000 is 72,000. One sixth of them cancelled their tickets. One sixth of 72,000 equals 12,000 $(\frac{1}{6} \times 72,000 = 12,000)$. 60,000 $(72,000 - 12,000 = 60,000)$ fans are attending this week.

11) The answer is $\frac{\sqrt{15}}{4}$

$\sin A = \frac{1}{4} \Rightarrow$ Since $\sin\theta = \frac{opposite}{hypotenuse}$, we have the following right triangle. Then:

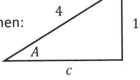

$c = \sqrt{4^2 - 1^2} = \sqrt{16 - 1} = \sqrt{15}, \cos A = \frac{\sqrt{15}}{4}$

12) The answer is 16π

The equation of a circle in standard form is: $(x - h)^2 + (y - k)^2 = r^2$, where r is the radius of the circle. In this circle the radius is 4. $r^2 = 16 \rightarrow r = 4$, $(x + 2)^2 + (y - 4)^2 = 16$

Area of a circle: $A = \pi r^2 = \pi(4)^2 = 16\pi$

13) The answer is 6.7×10^5

$670,000 = 6.7 \times 10^5$

14) The answer is 240

The ratio of boy to girls is $2:3$. Therefore, there are 2 boys out of 5 students. To find the answer, first divide the total number of students by 5, then multiply the result by 2.

$600 \div 5 = 120 \Rightarrow 120 \times 2 = 240$

15) The answer is 45

First, find the number. Let x be the number. Write the equation and solve for x. 150% of a number is 75, then: $1.5 \times x = 75 \Rightarrow x = 75 \div 1.5 = 50$, 90% of 50 is: $0.9 \times 50 = 45$

16) The answer is $\begin{bmatrix} -6 & 3 \\ 4 & -7 \end{bmatrix}$

First, find $2A$. $A = \begin{bmatrix} -1 & 2 \\ 1 & -2 \end{bmatrix}$; $2A = 2 \times \begin{bmatrix} -1 & 2 \\ 1 & -2 \end{bmatrix} = \begin{bmatrix} -2 & 4 \\ 2 & -4 \end{bmatrix}$

Now, solve for $2A - B$.

$2A - B = \begin{bmatrix} -2 & 4 \\ 2 & -4 \end{bmatrix} - \begin{bmatrix} 4 & 1 \\ -2 & 3 \end{bmatrix} = \begin{bmatrix} -2-4 & 4-1 \\ 2-(-2) & -4-3 \end{bmatrix} = \begin{bmatrix} -6 & 3 \\ 4 & -7 \end{bmatrix}$

17) The answer is $x \geq 5 \cup x \leq -1$

$x - 2 \geq 3 \rightarrow x \geq 3 + 2 \rightarrow x \geq 5$, Or $x - 2 \leq -3 \rightarrow x \leq -3 + 2 \rightarrow x \leq -1$

Then, solution is: $x \geq 5 \cup x \leq -1$

18) The answer is $\dfrac{8}{17}$

$\tan = \dfrac{opposite}{adjacent}$, and $\tan x = \dfrac{8}{15}$, therefore, the opposite side of the angle x is 8 and the adjacent side is 15. Let's draw the triangle.

Using Pythagorean theorem, we have: $a^2 + b^2 = c^2 \rightarrow 8^2 + 15^2 = c^2 \rightarrow$

$64 + 225 = c^2 \rightarrow c = 17$, $\sin x = \dfrac{opposite}{hypotenuse} = \dfrac{8}{17}$

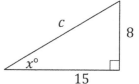

19) The answer is $x^{\frac{15}{4}}$

$$(x^6)^{\frac{5}{8}} = x^{6 \times \frac{5}{8}} = x^{\frac{30}{8}} = x^{\frac{15}{4}}$$

20) The answers are $0, -2, -4$

Frist factor the function: $f(x) = x^3 + 6x^2 + 8x = x(x + 4)(x + 2)$, To find the zeros, $f(x)$ should be zero. $f(x) = x(x + 4)(x + 2) = 0$, Therefore, the zeros are:

$x = 0, (x + 4) = 0 \Rightarrow x = -4, (x + 2) = 0 \Rightarrow x = -2$

21) The answer is 2

$\sin^2 a + \cos^2 a = 1$, then: $x + 1 = 3, x = 2$

22) The answer is $\frac{y}{6}$

Solve for x. $\sqrt{6x} = \sqrt{y}$. Square both sides of the equation:

$$\left(\sqrt{6x}\right)^2 = \left(\sqrt{y}\right)^2 \quad 6x = y; x = \frac{y}{6}$$

23) The answer is 61.28

$average = \dfrac{sum\ of\ terms}{number\ of\ terms}$, The sum of the weight of all girls is: $18 \times 60 = 1,080\ kg$

The sum of the weight of all boys is: $32 \times 62 = 1,984\ kg$, The sum of the weight of all students

is: $1,080 + 1,984 = 3,064\ kg$. $average = \dfrac{3,064}{50} = 61.28$

24) The answer is 36

Plug in the value of x and y. $x = 3$ and $y = -2$

$5(x - 2y) + (2 - x)^2 = 5(3 - 2(-2)) + (2 - 3)^2 = 5(3 + 4) + (-1)^2 = 35 + 1 = 36$

25) The answer is 15%

The question is this: 530.40 is what percent of 624?

Use percent formula: $part = \dfrac{percent}{100} \times whole$

$$530.40 = \frac{percent}{100} \times 624 \Rightarrow 530.40 = \frac{percent \times 624}{100} \Rightarrow 53,040 = percent \times 624 \Rightarrow$$

$$percent = \frac{53,040}{624} = 85. \, 530.40 \text{ is } 85\% \text{ of } 624.$$

Therefore, the discount is: $100\% - 85\% = 15\%$

26) The answer is $\frac{1}{2}$

The relationship among all sides of right triangle $30° - 60° - 90°$ is provided in the following triangle:

Sine of $30°$ equals to: $\frac{opposite}{hypotenuse} = \frac{x}{2x} = \frac{1}{2}$

27) The answer is $\frac{3}{5} + i$

To simplify the fraction, multiply both numerator and denominator by i.

$\frac{5-3i}{-5i} \times \frac{i}{i} = \frac{5i-3i^2}{-5i^2}, i^2 - 1,$ Then: $\frac{5i-3i^2}{-5i^2} = \frac{5i-3(-1)}{-5(-1)} = \frac{5i+3}{5} = \frac{5i}{5} + \frac{3}{5} = \frac{3}{5} + i$

28) The answer is 36

Let x be the smallest number. Then, these are the numbers: $x, x + 1, x + 2, x + 3, x + 4$

$$average = \frac{sum \ of \ terms}{number \ of \ terms} \Rightarrow 38 = \frac{x+(x+1)+(x+2)+(x+3)+(x+4)}{5} \Rightarrow 38 = \frac{5x+10}{5} \Rightarrow$$

$$190 = 5x + 10 \Rightarrow 180 = 5x \Rightarrow x = 36$$

29) The answer is $-\frac{1}{2}$

The equation of a line in slope intercept form is: $y = mx + b$. Solve for y.

$$4x - 2y = 12 \Rightarrow -2y = 12 - 4x \Rightarrow y = (12 - 4x) \div (-2) \Rightarrow y = 2x - 6$$

The slope is 2. The slope of the line perpendicular to this line is:

$$m_1 \times m_2 = -1 \Rightarrow 2 \times m_2 = -1 \Rightarrow m_2 = -\frac{1}{2}$$

30) The answer is $\frac{2}{x^4} + 2$

$$f\big(g(x)\big) = 2 \times \left(\frac{1}{x}\right)^4 + 2 = \frac{2}{x^4} + 2$$

ALEKS Mathematics Practice Test 2

1) The answer is $x^3 - 21x - 20$

Use FOIL (First, Out, In, Last), $(x - 5)(x^2 + 5x + 4) = x^3 + 5x^2 + 4x - 5x^2 - 25x - 20 = x^3 - 21x - 20$

2) The answer is -38

Use PEMDAS (order of operation):

$5 + 8 \times (-3) - [4 + 22 \times 5] \div 6 = 5 + 8 \times (-3) - [4 + 110] \div 6 =$

$5 + 8 \times (-3) - [114] \div 6 = 5 + (-24) - 19 = 5 - 43 = -38$

3) The answer is $\dfrac{-x-3}{2x^2-10}$

Simplify: $\dfrac{\frac{1}{2} - \frac{x+5}{4}}{\frac{x^2}{2} - \frac{5}{2}} = \dfrac{\frac{1}{2} - \frac{x+5}{4}}{\frac{x^2-5}{2}} = \dfrac{2\left(\frac{1}{2} - \frac{x+5}{4}\right)}{x^2-5} \Rightarrow$ Simplify: $\dfrac{1}{2} - \dfrac{x+5}{4} = \dfrac{-x-3}{4}$

then: $\dfrac{2\left(\frac{-x-3}{4}\right)}{x^2-5} = \dfrac{\frac{-x-3}{2}}{x^2-5} = \dfrac{-x-3}{2(x^2-5)} = \dfrac{-x-3}{2x^2-10}$

4) The answer is 78

Number of squares equal to: $\dfrac{52 \times 12}{4 \times 2} = 13 \times 6 = 78$

5) The answer is $x \leq 5$

Simplify: $5 + 2x \leq 15 \Rightarrow 2x \leq 15 - 5 \Rightarrow 2x \leq 10 \Rightarrow x \leq 5$

6) The answer is $\$635$

$55 \times \$66 = \$3,630$ Payable amount is: $\$4,265 - \$3,630 = \$635$

7) The answer is $x^{\frac{5}{2}}$

$(x^4)^{\frac{5}{8}} = x^{4 \times \frac{5}{8}} = x^{\frac{20}{8}} = x^{\frac{5}{2}}$

8) The answer is 2531.6

Underline the tenth place: 2531.$\underline{5}$8245, Look to the right if it is 5 or bigger, add 1 to the underlined digit. Then, round up the decimal to 2531.6

9) The answer is 125

$$25 \times \frac{x}{100} = 15 \Rightarrow 25 \times x = 1{,}500 \Rightarrow x = \frac{1{,}500}{25} = 60$$

Use percent formula: $part = \frac{percent}{100} \times whole$

$$25 = \frac{percent}{100} \times 20 \rightarrow 25 = \frac{percent \times 20}{100} \rightarrow 25 = \frac{percent \times 2}{10}, \text{ Multiply both sides by 10.}$$

10) $250 = percent \times 2$, **Divide both sides by 2:** $percent = 125$**The answer is**

9.48

$\$44 - \$34.52 = \$9.48$

11) The answer is $45°$

$90° + 45° = 135° \rightarrow 180° - 135° = 45°$

12) The answer is $(x - 4)(x - 3)$

$x^2 - 7x + 12 = (x - 4)(x - 3)$

13) The answer is $60 ft$

The relationship among all sides of special right triangle

$30° - 60° - 90°$ is provided in this triangle:

In this triangle, the opposite side of 30° angle is half of the hypotenuse.

Draw the shape of this question:

The latter is the hypotenuse. Therefore, the latter is 60 ft.

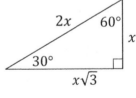

14) The answer is 5

$$C = \sqrt{(x_2 - x_1)^2 + (y_2 - y_1)^2}$$

$$C = \sqrt{(1 - (-2))^2 + (3 - 7)^2} \rightarrow C = \sqrt{(3)^2 + (-4)^2} \rightarrow C = \sqrt{9 + 16} \rightarrow C = \sqrt{25} = 5$$

15)The answer is 0.0143

$$\frac{2}{140} = \frac{1}{70} = 0.01428571429 \approx 0.0143$$

16)The answer is 12

$$\frac{a+b}{2} = 9 \Rightarrow a + b = 18, \frac{a+b+c}{3} = 10 \Rightarrow a + b + c = 30$$

$$18 + c = 30 \Rightarrow c = 30 - 18 = 12$$

17)The answer is $-\frac{1}{2}, -3$

$$x_{1,2} = \frac{-b \pm \sqrt{b^2 - 4ac}}{2a}, ax^2 + bx + c = 0 \Rightarrow 4x^2 + 14x + 6 = 0 \quad \Rightarrow \text{then: } a = 4, \ b = 14 \ \text{and}$$

$$c = 6$$

$$x = \frac{-14 + \sqrt{14^2 - 4\times4\times6}}{2\times4} = -\frac{1}{2}, x = \frac{-14 - \sqrt{14^2 - 4\times4\times6}}{2\times4} = -3$$

18)The answer is 13

$$7^5 \times 7^8 = 7^{5+8} = 7^{13} = 7^x \rightarrow x = 13$$

19)The answer is 33

$$33 + 36 + 30 = 99, Average = \frac{99}{3} = 33$$

20)The answer is $-1 + \sqrt{6}, -1 - \sqrt{6}$

$$x_{1,2} = \frac{-b \pm \sqrt{b^2 - 4ac}}{2a}, ax^2 + bx + c = 0, x^2 + 2x - 5 = 0 \Rightarrow$$

then: $a = 1, b = 2$ and $c = -5$

$$x = \frac{-2 + \sqrt{2^2 - 4\times1\times(-5)}}{2\times1} = -1 + \sqrt{6}, x = \frac{-2 - \sqrt{2^2 - 4\times1\times(-5)}}{2\times1} = -1 - \sqrt{6}$$

21)The answer is $(3, 3)$

$$\begin{cases} -2x - y = -9 \\ 5x - 2y = 9 \end{cases} \Rightarrow \text{Multiplication } (-2) \text{ in first equation} \Rightarrow \begin{cases} 4x + 2y = 18 \\ 5x - 2y = 9 \end{cases}$$

Add two equations together $\Rightarrow 9x = 27 \Rightarrow x = 3$ then: $y = 3$

22)The answer is 6

$|9 - (12 \div |2 - 6|)| = |9 - (12 \div |-4|)| = |9 - (12 \div 4)| = |9 - 3| = |6| = 6$

23)The answer is 32

$\log_2 x = 5$

Apply logarithm rule: $= \log_b(b^a)$, $5 = \log_2(2^5) = \log_2(32)$

$\log_2 x = \log_2(32)$, When the logs have the same base:

$\log_b(f(x)) = \log_b(g(x)) \Rightarrow f(x) = g(x)$, then: $x = 32$

24)The answer is $\dfrac{14}{x^3}$

$\dfrac{x^3}{14} \Rightarrow$ reciprocal is : $\dfrac{14}{x^3}$

25)The answer is 60

Plug in 140 for F and then solve for C.

$C = \dfrac{5}{9}(F - 32) \Rightarrow C = \dfrac{5}{9}(140 - 32) \Rightarrow C = \dfrac{5}{9}(108) = 60$

26)The answer is $12i - 54$

We know that: $i = \sqrt{-1} \Rightarrow i^2 = -1$

$(-3 + 9i)(3 + 5i) = -9 - 15i + 27i + 45i^2 = -9 + 12i - 45 = 12i - 54$

27)The answer is $-\sqrt{3}$

$\tan \dfrac{2\pi}{3} = \dfrac{\sin \frac{2\pi}{3}}{\cos \frac{2\pi}{3}} = \dfrac{\frac{\sqrt{3}}{2}}{-\frac{1}{2}} = -\sqrt{3}$

28)The answer is $28x + 6$

If $f(x) = 3x + 4(x + 1) + 2$, then find $f(4x)$ by substituting $4x$ for every x in the function. This gives: $f(4x) = 3(4x) + 4(4x + 1) + 2$

It simplifies to: $f(4x) = 3(4x) + 4(4x + 1) + 2 = 12x + 16x + 4 + 2 = 28x + 6$

29) The answer is $(4, -7)$, $\sqrt{3}$

$(x - h)^2 + (y - k)^2 = r^2 \Rightarrow$ center: (h, k) and radius: r

$(x - 4)^2 + (y + 7)^2 = 3 \Rightarrow$ center: $(4, -7)$ and radius: $\sqrt{3}$

30) The answer is $(x + 4)^2 + (y - 2)^2 = 1$

Use formula of a circle in the coordinate plane: $(x - h)^2 + (y - k)^2 = r^2 \Rightarrow$ center: (h, k) and radius: r, center: $(-4, 2) \Rightarrow h = -4, k = 2$

circumference $= 2\pi \Rightarrow$ circumference $= 2\pi r = 2\pi \Rightarrow r = 1$

$(x + 4)^2 + (y - 2)^2 = 1$

... So Much More Online!

Effortless Math Online ALEKS Math Center offers a complete study program, including the following:

✓ Step-by-step instructions on how to prepare for the ALEKS Math test

✓ Numerous ALEKS Math worksheets to help you measure your math skills

✓ Complete list of ALEKS Math formulas

✓ Video lessons for ALEKS Math topics

✓ Full-length ALEKS Math practice tests

✓ And much more...

No Registration Required.

Receive the PDF version of this book or get another FREE book!

Thank you for using our Book!

Do you LOVE this book?

Then, you can get the PDF version of this book or another book absolutely FREE!

Please email us at:

info@EffortlessMath.com

for details.

Author's Final Note

I hope you enjoyed reading this book. You've made it through the book! Great job!

First of all, thank you for purchasing this study guide. I know you could have picked any number of books to help you prepare for your ALEKS Math test, but you picked this book and for that I am extremely grateful.

It took me years to write this study guide for the ALEKS Math because I wanted to prepare a comprehensive ALEKS Math study guide to help test takers make the most effective use of their valuable time while preparing for the test.

After teaching and tutoring math courses for over a decade, I've gathered my personal notes and lessons to develop this study guide. It is my greatest hope that the lessons in this book could help you prepare for your test successfully.

If you have any questions, please contact me at reza@effortlessmath.com and I will be glad to assist. Your feedback will help me to greatly improve the quality of my books in the future and make this book even better. Furthermore, I expect that I have made a few minor errors somewhere in this study guide. If you think this to be the case, please let me know so I can fix the issue as soon as possible.

If you enjoyed this book and found some benefit in reading this, I'd like to hear from you and hope that you could take a quick minute to post a review on the book's Amazon page. To leave your valuable feedback, please visit: rb.gy/udoz88

Or scan this QR code.

I personally go over every single review, to make sure my books really are reaching out and helping students and test takers. Please help me help ALEKS Math test takers, by leaving a review!

I wish you all the best in your future success!

Reza Nazari

Math teacher and author

Made in the USA
Las Vegas, NV
29 April 2021